AGRICULTURE ISSUES AND POLICIES

DAIRY COOPERATIVES

PROFILES AND RESEARCH

AGRICULTURE ISSUES AND POLICIES

Additional books in this series can be found on Nova's website under the Series tab.

Additional e-books in this series can be found on Nova's website under the e-books tab.

AGRICULTURE ISSUES AND POLICIES

DAIRY COOPERATIVES

PROFILES AND RESEARCH

ANDREW WEST
AND
SKYE I. KELLY
EDITORS

Nova Science Publishers, Inc.
New York

For permission to use material from this book please contact us:
Telephone 631-231-7269; Fax 631-231-8175
Web Site: http://www.novapublishers.com

Library of Congress Cataloging-in-Publication Data

ISBN 978-1-62081-247-1

Published by Nova Science Publishers, Inc. †New York

CONTENTS

Preface **vii**

Chapter 1 Cooperative Theory, Practice, and Financing: A Dairy Cooperative Case Study **1**
K. Charles Ling

Chapter 2 Marketing Operations of Dairy Cooperatives, 2007 **49**
K. Charles Ling

Chapter 3 Measuring Performance of Dairy Cooperatives **71**
K. Charles Ling

Index **99**

PREFACE

Dairy cooperatives as a group are the most prominent among farmer cooperatives in terms of sales revenue and the important roles they play in the dairy industry, a major sector in agriculture. Their mission, functions, organization, governance, operations, market performance, financing, etc., are in full agreement with the economic theory of what cooperatives are and what cooperatives do. This book addresses some of the challenges faced by cooperatives and the issues raised by the alternative financing methods used. Classic literature on the economic theory of what cooperatives are and what they do in the market economy is discussed with a focus on measuring performance and marketing operations.

Chapter 1 – Cooperatives are the aggregates of economic units, such as farms. The cooperative is neither a horizontal integration of its member-farms nor a vertical integration between member-farms and the cooperative, but rather a third mode of organizing coordination. Cooperatives are owned, controlled, financed, and used by members for mutual benefits, with service at cost and proportionality being two basic principles. Farmers organize marketing cooperatives to access markets, exercise countervailing power *vis-à-vis* other market participants, promote competition, and thus enhance market efficiency. Cooperation as practiced by dairy farmers in marketing milk is an enduring business model that is in full accord with the economic theory of what cooperatives are and what cooperatives do. Members supply equity capital needed for the cooperative to carry out its core business of marketing members' milk. Capital financing, in general, is not a contentious issue for dairy cooperatives. For other cooperatives that have difficulties in raising capital from members, the issue is really a reflection of a certain gap between member purposes and cooperative functions. The solution lies in assessing what members want the cooperative to do and how much they are willing to finance it; the cooperative should operate accordingly for members' best interests. Social entrepreneurs have renewed interests in adopting cooperatives as an economic development tool to empower people to work toward their own economic destiny. Over the long term, cooperatives must be self-sustainable in order to be economically viable.

Chapter 2 – A total of 49,675 member-producers of the Nation's 155 dairy cooperatives marketed 152.5 billion pounds of milk, or 82.6 percent of all milk marketed, in 2007. Forty-five cooperatives operated 176 dairy processing and manufacturing plants, 12 handled milk through receiving stations only, and 98 had no milk handling facilities. Cooperatives marketed 71 percent of the Nation's butter, 96 percent of nonfat and skim milk powders, 26 percent of natural cheese, 7 percent of packaged fluid milk products, 4 percent of ice cream, 13 percent of ice cream mix, 11 percent of yogurt, 42 percent of dry whey products, 14 percent of sour cream, and 20 percent of condensed buttermilk. Cooperatives in the 1-billion-to-2-billion-pounds of milk group showed the most significant increase in the share of total cooperative milk volume, while the milk share of the larger sized groups, as a whole, was slightly lower. A total of 86 dairy cooperatives reported having 21,475 full-time and 2,944 part-time employees. Complete financial data submitted by 94 dairy cooperatives showed that total assets for the fiscal year were $8.41 per hundredweight (cwt); total liabilities, $6.09 per cwt; members' equity, $2.32 per cwt; and net margin before taxes, 28 cents per cwt, which represented a return on equity of 12.2 percent.

Chapter 3 – This report revises and simplifies the extra-value approach (developed previously in Research Report 166) for member-producers to evaluate their cooperative's performance. Extra value is defined as net savings after subtracting an interest charge on equity. For comparisons over time and among cooperatives, extra value is expressed as a percentage of operating capital to generate a scale-neutral and mode-neutral, extra-value index for each cooperative or group of cooperatives. The extra-value approach is also used to examine the performance of the surviving cooperatives following mergers and consolidations. The influence of cooperative size on performance is also examined.

In: Dairy Cooperatives Profiles and Research ISBN 978-1-62081-247-1
Editors: A. West and S. I. Kelly

Chapter 1

COOPERATIVE THEORY, PRACTICE, AND FINANCING: A DAIRY COOPERATIVE CASE STUDY*

K. Charles Ling

ABSTRACT

Cooperatives are the aggregates of economic units, such as farms. The cooperative is neither a horizontal integration of its member-farms nor a vertical integration between member-farms and the cooperative, but rather a third mode of organizing coordination. Cooperatives are owned, controlled, financed, and used by members for mutual benefits, with service at cost and proportionality being two basic principles. Farmers organize marketing cooperatives to access markets, exercise countervailing power *vis-à-vis* other market participants, promote competition, and thus enhance market efficiency. Cooperation as practiced by dairy farmers in marketing milk is an enduring business model that is in full accord with the economic theory of what cooperatives are and what cooperatives do. Members supply equity capital needed for the cooperative to carry out its core business of marketing members' milk. Capital financing, in general, is not a contentious issue for dairy cooperatives. For other cooperatives that have difficulties in raising

* This is an edited, reformatted and augmented version of United States Department of Agriculture, Rural Business– and Cooperative Programs, Research Report 221, dated April 2011.

capital from members, the issue is really a reflection of a certain gap between member purposes and cooperative functions. The solution lies in assessing what members want the cooperative to do and how much they are willing to finance it; the cooperative should operate accordingly for members' best interests. Social entrepreneurs have renewed interests in adopting cooperatives as an economic development tool to empower people to work toward their own economic destiny. Over the long term, cooperatives must be self-sustainable in order to be economically viable.

PREFACE

This study speaks to some recurrent issues regarding cooperative capital financing, such as those raised by participants at the USDA Rural Development Public Forum on Cooperative Research in 2005.

Cooperatives face many capital financing challenges. By the very nature of being a cooperative, its equity capital is provided by members and is limited by their financial means and their willingness to support the cooperative's undertakings. This may narrow the cooperative's scope of operation and expansion.

Furthermore, there are equity redemption issues. Members' equity capital held by the cooperative represents a substantial sum of their money. This competes with members' capital needs on the farm. Members tend to not favor a long equity revolving period, because the present value of revolved equity is diminished after being discounted.

When a cooperative's business is doing well, some members may perceive that its market valuation is higher than the book value and want to have access to the gain. This may force selling off the cooperative or converting it to a public corporation.

To overcome these challenges, alternative capital financing methods have been variously implemented, such as new-generation cooperative capitalization, issuing preferred stock, accumulating (unallocated) retained earnings, and changing State laws to allow non-member capital.

This report addresses some of the challenges faced by cooperatives and the issues raised by the alternative financing methods used. First, the classic literature on the economic theory of what cooperatives are and what they do in the market economy is reviewed. Then, a case study is presented on dairy cooperatives as a group to showcase the theory in practice. The case study serves as a knowledge base for answering questions regarding cooperative capital financing issues.

HIGHLIGHTS

Dairy cooperatives as a group are the most prominent among farmer cooperatives in terms of sales revenue and the important roles they play in the dairy industry, a major sector in agriculture. Their mission, functions, organization, governance, operations, market performance, financing, etc., are in full agreement with the economic theory of what cooperatives are and what cooperatives do (the classic *raison d'être* of cooperatives). The case study shows that financing the core business of marketing members' milk is generally not a contentious issue for dairy cooperatives. Capital financing could become an issue if a dairy cooperative embarks on ventures that are considered by members to be extraneous businesses.

Underpinning the case study are two epoch-marking treatises on the theory of the cooperative. "Economic Philosophy of Co-operation" by E. G. Nourse was the first academic paper on the theory of cooperation, published in the *American Economic Review.* Ivan Emelianoff's book, *Economic Theory of Cooperation: Economic Structure of Cooperative Organizations*, was the first work in the economic literature to give the cooperative its precise economic definition.

Nourse's economic philosophy of cooperation may be summed up in a nutshell: Cooperatives make it feasible for farmers to jointly market their products. The cooperative may evolve to a scale large enough to effectively bargain with other market participants and/or to avail itself of scale economies in processing and marketing operations. Subject to the same market disciplines and supply-demand-price dynamics as any business, the presence of the cooperative challenges other market participants to operate efficiently and thus strengthens the competitive market mechanism. When the market for members' products has become truly competitive, the cooperative may want to assume only a stand-by position but maintain the legal institutions and organizational capacity to re-enter the field, if necessary.

Emelianoff's theme was that for economic analysis of cooperatives, the economic structure of cooperative organizations should be clearly defined and that the definition should be free from the encumbrance of sociological, legal, technical, social-philosophical and ethical considerations. He established this definition: "Cooperative organizations represent the aggregates of economic units." In the agricultural context, farms are such economic units.

Being aggregates of member-farms, cooperative associations have these characteristics in common:

- The equity capital of a cooperative is the sum of advances needed for financing anticipated transactions of individual members of the cooperative.
- The member-owners of a cooperative are independent farmers who have chosen to coordinate certain activities via a cooperative.
- The surplus or deficit of a cooperative is the account payable to, or receivable from, the member-patrons of the cooperative on their current transactions.
- The sum for patronage refunds to members is the sum either underpaid (overcharged) to members, or—in case of a deficit—overpaid (undercharged) to members on their transactions through the marketing (or purchasing) cooperative.
- The dividend on capital, if any, is the interest payment for using capital advanced by members, not a distribution of income.
- All the economic functions of a cooperative are ultimately the economic functions of the member-farms performed through the cooperative as their collective branch. Therefore, all economic services of cooperative association are performed at cost.

The theory as expounded by Nourse is from the perspective of market performance of cooperatives. Emelianoff's thesis delineates the economic structure of cooperatives and the governance, financial, and functional corollaries. Market performance and economic structure of dairy cooperatives are shown to be as prescribed by Nourse and Emelianoff. (Table 2 in this report (page 11) summarizes their salient points and makes a side-by-side comparison to dairy cooperative practice.) Cooperation as practiced by dairy farmers in marketing milk is an enduring business model that is in full accord with economic theory.

Just like any other business firm, dairy cooperatives require an adequate level of capital to market members' milk. Besides bargaining/negotiating for milk prices and terms of trade, they may own and operate milk-handling facilities, do value-added processing, and/or provide milk marketing-related and other member services, as the case may be. However, member equities are the source of capital of dairy cooperatives. A cooperative's ability to raise capital is thus constrained by members' financial resources. Equities of dairy cooperatives comprise common stock, preferred stock, retained earnings, and allocated equities (such as retained patronage refunds, capital retains, base capital, etc.)

Members are usually supportive of the financing need if the capital requirement is for the cooperative to carry out the milk marketing functions they want it to perform. A cooperative may face financing issues if it attempts to invest in extraneous businesses unless the members are convinced that the new ventures would:

- solidify the market for members' milk, or
- help market members' milk, or
- add value to members' milk, and
- benefit members the most among all available alternatives of investing the capital.

Some dairy cooperatives have tried alternative equity financing methods to leverage cooperative members' capital commitment. They have tried structuring subsidiaries as public stock corporations or as limited liability companies, having joint ventures with other firms or organizing as a new-generation cooperative. A few have issued preferred stock for quite some time, mostly to members.

The solution to capital financing issues lies in assessing what members want the cooperative to do and whether they are willing to adequately finance it. A cooperative must be self-sustainable in order to be economically viable over the long term.

The experience learned from the case study of dairy cooperatives provides some answers to the frequently asked questions concerning cooperative capital financing, such as those raised by presenters at the USDA Rural Development Public Forum on Cooperative Research in 2005. The issues are addressed by individual subject matter in the table below.

Financing issues	Some abridged answers
Member equities: Cooperative equities are furnished by members and therefore are limited. How does a cooperative gain access to capital without incurring long-term debt, with- out selling off the cooperative, or without going public:	Members organize or join a cooperative to mar-ket their farm production. They should provide the cooperative with capital at a level that is commensurate with the functions they want the cooperative to perform and the benefits they want to derive from it.

(Continued)

Financing issues	Some abridged answers
• When the cooperative needs more capital? • When members agitate to gain access to the perceived high market value of the business? • When pressure mounts to shorten the equity revolving period?	If market value of the cooperative is higher than the book value, it means the cooperative's earnings and potential future earnings are higher than can be expected, given its level of equity capital. Members gain access to this higher earning ability by receiving higher pay prices, premiums, and patronage refunds. Selling off the cooperative to gain the value of the business is tantamount to "killing the goose that lays the golden egg." Eventually it is up to members to decide if they want the cooperative to be viable or if they prefer other alternatives.
New-generation cooperatives: Are they the answer to cooperative financing issues?	A new-generation cooperative requires members to pay equity up front to acquire the delivery rights. This cooperative model has its own issues, especially those concerning delivery rights and property rights. In addition, many new-generation cooperatives are organized for business opportunities that resemble venture-capital investment. They tend to process one product or a narrow range of products. This presents additional risks as compared with a cooperative that is organized to market members' product(s) through a variety of marketing channels.
Preferred stock: What effects might issuing preferred stock have on a cooperative's practice?	The effects preferred stock may have on a cooperative's practice depend on what rights are specified. Preferred stock that pays dividends and has preference in assets over common stock in the event of the dissolution of the cooperative—the most common type of preferred stock—probably would not have any impact.
Retained earnings: Many cooperatives expand non-member busi- nesses to accumulate retained earnings as perma- nent equity.	Cooperatives may have non-member business for various reasons. In any case, retained earnings belong to the cooperative and thus are jointly owned by members.

Financing issues	Some abridged answers
What might be the long-term effects of this practice on governance?	However, a cooperative would not be conforming to the Capper-Volstead Act requirement if its non-member business were to exceed 50 percent of total sales.
Outside (non-member) capital: What changes in governance, organizational structure, and practice may be brought about by the new cooperative laws enacted by some States that allow outside equity capital?	There is a large variation regarding voting power and earning distribution, etc., among the few State laws that allow cooperatives to have investors. Differences in governance and earning distribution rules will influence cooperative organizational structure and practice. It is better to analyze them on a case-by-case basis. Furthermore, not every cooperative newly incorporated under these State laws has investors.
Untapped equity in rural America: How plausible is the contention that there is a lot of untapped equity in rural America that cooperatives just do not have access to and that should be allowed to be invested in cooperatives?	If the farm sector equity is any indication, it is not clear how much of it is untapped or available for off-farm investment. The equity-to-asset ratio of the farm sector is between 88 and 90 percent during the 5-year period 2006-2010. About 96 percent of the assets are in land and real estate and farming assets. Only around 4 percent of the assets are in financial assets, and they may not be available for off-farm investing. No comparable financial data for rural America is readily available, but recent assessment of the state of the rural economy shows that it faces significant challenges. It is highly unlikely that there is a lot of untapped equity to be found in the rural area.

INTRODUCTION

Cooperatives need capital in their normal course of operation. Furthermore, marketing value-added products and the capital-intensive nature of modern operations also bring up financing issues. In addition, modern technology is usually embodied in new plants and equipment that require a large volume of throughput to achieve economies of scale and also a large sum of capital investment.

There is always an internal tension between a cooperative's need for equity capital to carry out its functions and the members' need for financing their farming operations and for living expenses. It is not unusual to hear laments that “our cooperative is not able to leverage its strong brand for potentially lucrative market growth due to the constraints of member equity financing.”

Indeed, cooperative financing was prominently mentioned as one of the major subjects that require research attention by most presenters at the USDA Rural Development Public Forum on Cooperative Research (*U.S. Department of Agriculture, 2005*). The following list summarizes the cooperative financing issues raised at the forum:

1) The chief obstacle facing farmer cooperatives is the lack of access to sufficient capital to fund their organizations in this new, expanded and globalized marketplace. Because equity capital is furnished by members and therefore is rather limited, how does a cooperative gain access to capital without incurring long-term debt?
2) When a cooperative is successful (profitable), its perceived market valuation may be worth more than the book value. This may agitate some members to demand cashing in the appreciated value. How does a cooperative accommodate such demand and provide a vehicle for members to gain access to the value of the business without selling the cooperative or going public?
3) Some cooperatives face equity redemption issues. Others have attempted to restructure the cooperative in order to have more room for raising capital. Are there alternatives short of selling the cooperative or going public?
4) The advent of new-generation cooperatives that would confer delivery rights commensurate with tradable membership shares seems to offer a promising alternative for financing cooperatives. Are new-generation cooperatives the answer?
5) One of the ways to access capital is by issuing preferred stock. What effects might this have on a cooperative's practice?
6) Many cooperatives are expanding non-member business to generate profits that can be held in permanent equity within the cooperative. What might be the long-term effects of expanded non-member business on cooperative governance?

7) Cooperative laws in some States have been changed to allow outside (non-member) equity capital. What changes in governance, organizational structure, and practice may be brought about by these new laws?
8) Some contended that there is a large amount of untapped equity in rural America that cooperatives do not have access to and that should be allowed to be invested in cooperatives. How plausible is this contention?

To set the stage for addressing these cooperative financing issues, it is useful to refresh our understanding of the cooperative basics—the unique characteristics of the cooperative form of business: what cooperatives are, what they do, how they do it, what their role is in terms of market performance, and how their operations are financed. To show the relevance of cooperative theory in practice, dairy cooperatives are used as a case study. The lessons learned will provide some useful insights for answering the above questions.

ECONOMIC THEORY OF THE COOPERATIVE

For reviewing cooperative basics, the narratives of two epoch-marking treatises on the economic theory of the cooperative are presented: one on the cooperative's roles in the marketplace and the other on the economic structure of the cooperative.

"Economic Philosophy of Co-operation" by E. G. Nourse was the first academic paper on the theory of cooperation, published in the *American Economic Review (Nourse, 1922; Hess*). This piece, supplemented by the oft-quoted "brief remarks" he made years later (*Nourse, 1945*), is the basis of the first narrative.

The second narrative is drawn from the ideas of a book by Ivan Emelianoff, *Economic Theory of Cooperation: Economic Structure of Cooperative Organizations*. This book was the first in the economic literature to give the cooperative its precise economic definition, and his work marked the beginning of a new era in the development and evolution of cooperative theory.

Cooperative Basics I: Cooperation for market efficiency

Nourse's primary focus was on the role agricultural cooperatives played in the marketplace. This arose from his observation that the attempt to apply the cooperative form of organization to economic needs and problems in agriculture was critically important.

Purposes of cooperation

The following examples are taken from Nourse's paper to illustrate how farmers organize cooperatives to perform various market functions jointly and efficiently in various market situations—functions that cannot be satisfactorily carried out alone by individual farmers:

1) Cooperation for market access—An example is a small fruit-producing area far from any large market. The product is perishable, hence both risk and marketing expense are high. Total product volume is not large enough to attract a private distributor. Facing this situation, producers have the option of organizing a cooperative association to market their products. These cooperatives have frequently demonstrated the ability to achieve successful results where private, outside entrepreneurship fails to perform.
2) Local to regional coordination—A local cooperative creamery may initially be effective in meeting the competition of other small, private creamery operations. However, when competing creameries have grown to be entities of great size, the competition must be met by a distributing organization of equal scope. This will often be achieved through a federation of the cooperative creameries across a region which may embrace an entire State, several States or parts of a State.
3) Region-wide associations—In many instances, growers in horticultural regions have organized and integrated highly efficient businesses that serve producers across an entire production region by assembling, processing, and distributing their products. These agencies have eliminated wasteful competition both at the local shipping point and at the central markets. Furthermore, they are the instruments of the producer and owner of the goods, and hence are likely to be more aggressive in the effort to reduce expense and wastage in the handling process and to improve quality and enlarge outlets.

(Author's note: Cooperative organizations covering entire production regions have been most prevalent in California because of the characteristics of the State's economic geography. This type of cooperative organization was called "the California plan" and was promoted on a national scale in the 1920s by Aaron Sapiro, with varying degrees of success and failure (*Sapiro; Larsen, et al.).*

Countervailing Power

The above examples show how cooperatives are organized and grow to enable farmers to exercise "countervailing power" in the marketplace, although the term was not coined until the 1950s, when economist John Kenneth Galbraith cited the type of cooperatives made famous by Sapiro as an example for his explanation.

Nourse certainly recognized the importance of countervailing power if cooperatives are to have a strong market position. As he stated: "Possibly the keynote of the philosophy lies in the idea that a means must be found for giving agriculture a type of organization whose productive and bargaining units respectively will expand in step with the growing needs of the agricultural techniques (and its accompanying capital demands) and of the size requisite to an effective bargaining position in contact with the units of commercial organization with which they must deal."

Pro-Market

Nourse said that the theoretical implication of agricultural cooperation "is preeminently that of functional reorganization rather than comprehensive economic regeneration." In other words, the farmer takes the essential facts of the market as given and, working together with other producers through the cooperative, seeks to be in the most effective market position to compete. Thus, the distinctive economic philosophy of this business form is viewed "as a means to improve the lot of both farmer and consumer by improving the efficiency of the economic machine."

Cooperatives enable farmers to effectively compete in the marketplace and garner market signals that put them in a position of prompt and sensitive response to the reaction of the consuming public and guide their farming business decisions. According to Nourse, the cooperative objective is twofold (*Nourse, 1945*):

1) "It is to make the most economical and efficient market channel by which whatever volume of product farmers see fit to produce gains

access to the attention and the purchasing power of all who might use such a product. (For supply-buying cooperatives, most economical access to the best sources of the goods they need.) Thus, a true supply-anddemand price is allowed (and aided) to express itself for the guidance of producers."

2) "It aims to reflect these market conditions back most promptly and fully to producers in ways that will both guide and, so far as possible, assist them in changing their methods so as to continue production and to prosper or to shift to more suitable lines of production."

Competitive Yardstick

In Nourse's view, the cooperative is a means for promoting and maintaining competition in the marketplace. The supply-demandprice dynamic "provides a powerful stimulus to the association to devise further economies of method which will enable them to maintain the level of net returns to the grower. Such competition also spurs the private agency to outdo the cooperative in its efficiency in order to hold its business."

He used the term "yardstick" years later (*Nourse, 1945*), when he said the place for the agricultural cooperative in the Nation's business "is primarily that of 'pilot plant' and 'yardstick' operation. Its objective is not to supersede other forms of business but to see that they are kept truly competitive."

The cooperative is to "occupy certain strategic points, and there to set a plane or pace of competition which will assure for the farmer efficient service at true long-run cost." When such services (manufacturing, distributing, transporting, financing, etc.) are furnished efficiently and economically (which means in a truly competitive manner), "there is no occasion for the farmer to occupy the field and divert some of his capital and some of his managerial time and effort to these tasks and away from his main enterprise of farm production."

Farmers should remain vigilant. Nourse cautioned: "It is of the upmost importance, however, that farmers shall have both the legal institutions and the organizational 'know-how' to step into these fields when and to the extent that service is inadequate or unduly high in cost. It is important also that they remain in each of these fields with an organization sufficiently large to attain high efficiency so that farmers shall be protected against any subsequent lapse in the quality of service or temptation to profiteer in charges by the non-cooperative service agencies.

"But it is just as important that the cooperatives recognize when they have in fact attained their real objective by demonstrating a superior method of processing or distribution or by breaking a monopolistic bottleneck, and that they should then be content merely to maintain 'stand-by' capacity or a 'yardstick' operational position rather than try to occupy the whole field or a dominating position within it. In some cases, they may be well advised in entirely terminating operations once they have stimulated regular commercial or manufacturing agencies to competition amongst themselves."

Nourse's economic philosophy of cooperation may be summed up in a nutshell: Cooperatives make it feasible for farmers to jointly market their products. The cooperative may evolve to a scale large enough to effectively bargain with other market participants and/or to avail itself of scale economies in processing and marketing operations. Subject to the same market disciplines and supply-demand-price dynamics as any business, the presence of the cooperative challenges other market participants to operate efficiently and thus strengthens the competitive market mechanism. When the market for members' products has become truly competitive, the cooperative may want to assume only a stand-by position but maintain the legal institutions and organizational capacity to reenter the field, if necessary.

Cooperative Basics II: Cooperatives Are the Aggregates of Economic Units

While there had been many references regarding cooperative principles, the question "what is a cooperative?" was never clearly answered until Ivan Emelianoff took up the task to do it.

In *Economic Theory of Cooperation: Economic Structure of Cooperative Organizations*, Emelianoff carefully reviewed the worldwide literature on cooperative theory from the late 19th century until 1939. He came to the conclusion that for economic analysis of cooperatives, the economic structure of cooperative organizations should be clearly defined and that the definition should be free from the encumbrance of sociological, legal, technical, social-philosophical, and ethical considerations.

Against this backdrop, Emelianoff established this definition: "Cooperative organizations represent the aggregates of economic units." This "bare bones" definition crystallizes the essence of what cooperatives should have in common.

"Aggregate" is commonly defined as: "Any total or whole considered with reference to its constituent parts; an assemblage or group of distinct particulars massed together" (*American Heritage Dictionary*). Further, as defined by

Emelianoff: "An economic unit, or economic individual, is an economic body admittedly complete and sufficiently integrated for individual existence and independent (in conditions of an exchange economy—interdependent) economic functioning."

In the agricultural context, farms are such economic units. The nature of cooperative associations as aggregates of member-farms is clearly discernible in the embryonic forms of such associations. For example, a buying club of farmers may want to purchase certain goods together, such as fertilizer.

The buying club would have someone take orders from member-farmers and place orders with a vendor, as well as perform other related chores. If the vendor requires a deposit, members may advance money to the buying club for the deposit requirement in proportion to their respective buying volume.

There may be an elected committee to facilitate decisionmaking if the number of members is large. Each member may have one vote if their purchasing volumes are about the same. Otherwise, some forms of proportional voting may be adopted to conciliate large-volume members.

When the fertilizer (for example) is delivered, members pay the balance of their obligations. After the transactions have been completed, payment to the vendor and other expenses are subtracted from the sum of money paid by members. Any surplus is returned to members in proportion to the volume of fertilizer they have purchased.

This buying service is conducted at cost; every aspect of a member's transaction through the buying club is in proportion to their patronage (buying) volume. The buying club may be disbanded after fulfilling its joint-buying purpose.

This scenario shows that the buying club represents the aggregate of its member-farms, through which they purchase fertilizer. If the buying club metamorphoses into a permanent purchasing cooperative association, the picture may look more complicated. However, the underlying nature of the cooperative as an aggregate of member-farms remains the same.

In this new scenario (i.e., a permanent purchasing cooperative), the person who manages buying orders and other chores will be the manager of the cooperative (usually a hired professional). The committee of members becomes the board of directors. Advanced payments by members to the cooperative become equity capital for financing the operation and for carrying inventories and owning facilities.

Year-end surplus is returned to members as refunds in proportion to patronage volume, but a portion may be retained as revolving capital. The principles of proportionality and service at-cost remain intact, but their practices may be less evident because the operation has become more complex.

Although the above example is based on purchasing cooperatives, the same line of reasoning also applies to marketing cooperatives. The difference between purchasing and marketing cooperatives is: instead of procuring goods, a marketing cooperative markets products produced by member-farms.

In either case, the member-farms coordinate their activities through the cooperative, but each fully retains its economic individuality and independence.

A cooperative may be described as a center of member-patrons' coordinated activities, or as an agency owned and controlled by members through which they conduct their business. In this respect, it is identical with the special departments or branches of individual member-farms.

For example, a dairy cooperative is the collective marketing arm of its member dairy farms; a farm supply cooperative is their supply purchasing department; and a livestock-genetics cooperative is the breeding service branch for its members. As some would say, "A cooperative is an off-farm extension of the farming business."

Being aggregates of member-farms, cooperative associations have these characteristics in common:

- The equity capital of a cooperative is the sum of advances needed for financing anticipated transactions of individual members of the cooperative; it is not the same as the entrepreneurial capital of an investor-owned corporation.
- The member-owners of a cooperative are independent farmers who have chosen to coordinate certain activities via a cooperative. They are not the same as the stockholders of an investor-owned corporation, who are a diverse set of shareholders joined solely by common investment.
- The surplus or deficit of a cooperative is the account payable to, or receivable from, the member-patrons of the cooperative on their current transactions; this is not the same as the profit or loss of an investor-owned corporation.
- The sum for patronage refunds to members is the sum either underpaid (overcharged) to members, or — in case of a deficit — overpaid (undercharged) to members on their transactions through the marketing (or purchasing) cooperative; the sum for patronage refunds is not the profit of the cooperative or its income.

- The dividend on capital, if any, does not represent a profit or any income of the cooperative; it is the interest payment for using capital advanced by members. By contrast, investor-owned corporations pay dividends to shareholders out of earnings.
- All the economic functions of a cooperative are ultimately the economic functions of the member-farms performed through the cooperative as their collective branch or collective department. Therefore, all economic services of a cooperative association are performed at cost.

Emelianoff emphasizes: "None of such traits can be unreservedly used as an unerring test of a truly cooperative organization, since these traits only indirectly disclose the economic character of the cooperative aggregate...The only comprehensive and indisputable test of the cooperative character of organizations is their aggregate structure."

The unique aspects of cooperative character, however, are often not readily apparent. There are many reasons for this, some examples being:

- Cooperatives only reflect the characters and aspirations of their membership, which are diverse and manifest the diversity of the population, the geographical regions and the commodities involved. Such differences directly, or indirectly, have a certain bearing on the character of an association and its cooperative ideals. The variability of the external characteristics of cooperatives is kaleidoscopic and infinite. Differences in their external and superficial features obscure cooperatives' ultimate economic character of being aggregates of their member-farms.
- Most cooperatives are incorporated. The legal vestments of incorporated cooperative associations also cloak their economic structure as aggregates of member-farms to such a degree that they are often mistaken to be the same as investor-owned corporations. This is one of the principal sources of confusion in understanding cooperative organizations.
- A lack of distinction between the concept of an investor-owned corporation as a profit-seeking economic unit and the concept of a cooperative as an agency of its member farms is another factor that confuses many. Use of common accounting terminology for both business models adds to this confusion. As the above list of cooperative characteristics shows, such conventional terms as

"profit," "capital," "shareholders," "dividends," etc., should be used with reservations when describing cooperatives.
- In governance, a cooperative board of directors — including its board election rules, composition, function, responsibilities, and interaction with management — is not the same as the board of an investor-owned corporation (especially the publicly traded ones). Consequently, the role of the top manager of a cooperative is also somewhat different from that of an investor-owned corporation (even if they have the same title).

Emelianoff's definition that cooperative organizations represent the aggregates of associated economic units provides a clear insight into how cooperatives organize and function.

In a paper dealing with the issue of economic coordination some 45 years later, James Shaffer echoed (though without citing) Emelianoff's definition of cooperatives as aggregates of member-farms. Because member-farms are independent entities, represent independent profit centers, and act independently (except that they jointly own the cooperative), the cooperative association is neither a horizontal integration of its member-farms nor a vertical integration between member-farms and the cooperative. He asserted that "the cooperative is a third mode of organizing coordination."

Summary of Cooperative Basics

The first narrative on the cooperative basics, following Nourse's work, is from the perspective of market performance of cooperatives. To sum up:

- Cooperatives are organized for efficiently carrying out specific business functions for member-producers.
- Cooperatives can be of any size (and can be local, regional, or national in scope) that allows them to function efficiently in the marketplace.
- Cooperatives afford farmers the organizational sizes that are necessary for exercising countervailing power to effectively deal with other market participants.
- Cooperatives are pro-market; they let the market supply-and-demand price be the guidance for producers.

- Cooperatives are a means for farmers to promote and maintain competition—as the competitive yardstick.
- In those fields where the market has become truly competitive and farmers can be well served by other firms, cooperatives may want to cede the field and assume only a stand-by position (to preserve members' capital, time, and efforts for use on the farm), while maintaining the legal institutions and organizational capacity to step in if there is a relapse of market inadequacy.

The second narrative, based on Emelianoff's theory, delineates the economic structure of cooperatives and the governance, financial, and functional corollaries. In summary:

- Cooperative organizations represent the aggregates of economic units.
- A cooperative is an agency owned and controlled by members through which they conduct their business.
- Each member-farm fully retains its economic individuality and independence.
- The board of directors is elected from among member-farmers.
- Proportionality and service at-cost are two basic principles.
- Members provide advances (i.e., equity capital) for financing the cooperative.
- The surplus or deficit of a cooperative is the account payable to, or receivable from, the member-patrons.
- Patronage refunds are the money returned to members who have been underpaid or overcharged.
- Dividend on capital, if any, is interest payment for using members' capital.
- Being an aggregate of member-farms, the cooperative is neither a horizontal integration of its members nor a vertical integration between the cooperative and its members. It is a third mode of organizing coordination.

Together the two narratives constitute a comprehensive framework for understanding cooperatives— what they are, what they do, how they do it, and their role in the marketplace. In the next section, dairy cooperatives are examined to relate their practice to the theoretical framework and illustrate how well the theory fits the reality, and *vice versa*.

Cooperative Practice— Dairy Cooperatives as a Case Study

Dairy cooperatives' sales of milk and dairy products represented 42 percent of total commodity marketing by all agricultural cooperatives in the United States in 2007, making them, as a group, the most prominent among all agricultural marketing cooperatives (*Deville, et al.*).

Dairy cooperatives also occupy major market shares within the dairy industry, especially at the first-handler level and in the manufacture of "hard" dairy products (butter, cheese, and milk powders). In 2007, there were 155 dairy cooperatives in the nation. A total of 49,675 member-producers, or 84 percent of the nation's licensed dairy farms, delivered 152.5 billion pounds of milk, or 83 percent of all milk marketed (*Ling, 2009*).

Cooperatives marketed 71 percent of the nation's butter, 96 percent of nonfat and skim milk powders, 26 percent of natural cheese, and 42 percent of dry whey products. Their shares of "soft" and cultured products were less significant: 4 percent of ice cream, 13 percent of ice cream mix, 11 percent of yogurt, and 14 percent of sour cream. They processed 7 percent of the Nation's packaged fluid milk products in 2007.

Mission and Functions

Dairy farmers organize cooperatives to jointly and efficiently market their milk. Milk is a "flow" product (cows are milked twice or thrice daily) and is highly perishable; it must be picked up from the farm and delivered to the market (milk plants) soon after it is produced. By working together through their cooperatives, farmers want to have better control over the movement of the milk through the marketing channel and attain higher value for the milk.

The functions and services the farmers demand of their respective cooperatives vary, depending on the specific market situation the members of a cooperative face and their particular needs. Dairy cooperatives may be charged by members with the responsibility of performing one or more (or all) of the following marketing functions:

- Provide an assured market — Typically there is a written or tacit agreement between a member and the cooperative that the cooperative is the exclusive marketing agent of the member's milk.

- Negotiate milk pay price and terms of trade with milk buyers (investors-owned processors).
- Collect and ensure payment from milk buyers.
- Check weights and tests — To ensure that milk payment a member receives is accurate and commensurate with the quantity and quality of the milk delivered.
- Arrange for milk hauling — Arrangement must be made to have milk picked up from the farm in a timely fashion and delivered to the plant of first-receipt. This can be performed by the cooperative's own haulers, by contract haulers or by haulers retained by members. The cooperative may also be responsible for setting or negotiating hauling rates.
- Provide field services — Cooperatives typically have field service personnel to assist with on-farm production problems and regulatory and inspection issues for the farm to achieve quality milk production.
- Disseminate market information — Information on the situation and outlook of the milk market is provided to members for use in making dairy farming business decisions.
- Other marketing-related services that help members deal with all the minutiae related to producing and marketing quality milk.

In addition, dairy farmers may ask their cooperative to leverage its group strength to procure various other services to help sustain their farming operations and farm life. Some of the services may be for the cooperative to offer or help arrange:

- Milking supplies and equipment or farm supplies.
- Insurance products — Such as disaster insurance for the farm, health and/or life insurances for the farmer and the farm employees and their families, and farm workers' compensation.
- Retirement programs.
- Risk management services to deal with market uncertainties.
- Farm business consulting services — such as farm expansion feasibility studies and business plans.
- Operating capital and facility capital financing.
- Financial planning services.
- Livestock marketing services — mainly for culled cows and calves.
- Other services that may help members' farming operations.

Organization

Dairy cooperatives can be of any size (and can be local, regional, or national in scope), depending on whatever scale the membership considers to be the most appropriate for marketing their milk.

A small local cooperative may have a few member-farms and market less than 1 million pounds of milk a year. A regional one may have hundreds or thousands of members in more than one State with milk pounds in the millions or even billions. The Nation's largest dairy cooperative has about 10,000 member-farms that spread over all 48 contiguous States and together deliver tens of billions of pounds of milk.

All dairy cooperatives are known to be centralized organizations with direct membership. A limited number may have other dairy cooperatives as association members, but the practice is usually for accommodating the fact that the cooperative is the marketing agent of all or part of the milk, dairy products, or services of these association members.

Dairy cooperatives operating in the same market may form marketing agencies in-common to rationalize milk hauling and shipment for reducing transportation costs, to share market information, or to collectively bargain with buyers for higher prices for milk or dairy products marketed.

Governance

Members of dairy cooperatives exercise ownership and business controls through a board of directors that is elected from among member-farmers. Candidates for the board are typically nominated by a committee of elected members who are not directors. Elections of the directors are usually done at the annual membership meeting.

If a cooperative is large, in terms of membership or geographical area, members may be grouped into districts (or areas/regions/divisions/locals, as the case may be). Then the directors may be nominated from the district and elected at the cooperative's annual meeting. Districts are usually drawn such that members in the same district are more or less homogeneous, and voting at the district level is typically by one member/one vote. The number of directors each district is entitled to may be different due to proportionality considerations based on milk volume. Some boards may have at-large members.

Also in a large cooperative, a delegate body elected by members may be needed to channel information and make decisions on behalf of the membership. The delegate body may be empowered to represent the membership in all decisions, except for matters that specifically require votes by the entire membership.

A limited number of dairy cooperatives are known to have non-member directors, typically in the States where they are required by law. Non-member directors usually play an advisory, non-voting role on the board.

An executive committee of elected officers and selected board members may be constituted to facilitate decisionmaking when the board is not in session. The board may also appoint several committees to carry out specific board functions, such as audits, finance, membership, and marketing committees.

The board controls the cooperative's business on behalf of members and makes major decisions, sets the policy, and determines the overall direction of the cooperative for the management to carry out in its day-to-day operations. The separation of the responsibility of the board (governance) and the role of management (managing) is often emphasized. Another very important function of cooperative board members is serving as a conduit of communication between the management and the rank-and-file members.

Operations

Dairy cooperatives perform various marketing functions to carry out the most important task of providing an assured market for members' milk. They may engage in one or more of these activities:

- Bargaining—Find a market for members' milk and bargain/negotiate with milk buyers for milk prices and terms of trade.
- Fluid processing—Own or retain plant capacity to process some or all member milk into fluid products. Fluid plants may also process soft and cultured products.
- Niche marketing—Own or retain plant capacity to process some or all member milk into specialty (niche) products.
- Making hard products—Own or retain plant capacity to manufacture hard dairy products. Manufacturing plants also provide a home for milk when it is in excess of market demand and transform the milk into storable products for further processing or later distribution.

Of the 155 U.S. dairy cooperatives, 108 may be classified as bargaining cooperatives because bargaining was their only, or main, marketing activity. Four were fluid processing cooperatives whose business was predominantly in processing and distributing fluid products. Nineteen were niche marketing cooperatives. The remaining 24 may be called diversified cooperatives, having bargaining and one or more processing/manufacturing functions as their main operations.

Besides assuring a market for members' milk, dairy cooperatives may also perform some or all of the other milk marketing functions listed in the mission and functions section above. In addition, they may procure farm supplies or provide other services for members.

Dairy cooperatives also provide services to milk buyers in accordance with the terms of trade negotiated, such as delivering milk on schedule, maintaining quality control and related laboratory services, preconditioning or standardizing milk, and/or fulfilling full-supply contracts.

Market Performance

A cooperative affords dairy farmers the organizational size that is necessary for exercising countervailing power to effectively bargain and deal with milk buyers and other market participants.

The dairy industry has evolved such that dairy cooperatives and processors have developed into what may be characterized as symbiotic relationships, and there is a high degree of "division of labor."

Because dairy cooperatives are organizations of farmers, they have the comparative advantages of working closely with members for assembling milk, providing field services, and performing farm-related functions. It is these advantages that accord them the predominant market share at the first-handler level.

Along with this dominance in milk procurement is the responsibility of balancing milk supply. Many dairy cooperatives maintain plant capacity to manufacture reserve and surplus milk into storable products like butter, milk powders, and cheese. Consequently, they have major market shares of these hard products. Like a reservoir, these cooperative plants absorb milk in excess of demand and provide supplemental milk to the market when it is needed.

Many processors rely on dairy cooperatives for milk supplies that are tailored to their requirements such as volume, quality, composition, and delivery schedule—what are called full-supply contracts—so they can focus their attention on the sectors where they are dominant: making fluid, cultured, and soft products (and lately cheese) and further processing and packaging dairy products for the consumer market. These sectors tend to be capital-, technology-, and service-intensive and are exposed to high product and market risks.

Farmers, who are generally risk-averse and have many demands on their financial resources on the farm, probably prefer to stay out of these sectors rather than compete head-on with processors (their milk customers), as long as the market performs well and their farming business can be sustained.

Still, there is a substantial number of dairy cooperatives operating in these sectors, although as a whole their market share is not high. The upshot is that although dairy cooperatives are generally less active in these sectors, they have the size, organization, and wherewithal to enter the market if the situation calls for it.

As far as dairy cooperatives are concerned, Nourse's prescription regarding market performance of cooperatives still fits the reality very well.

Financing

Based on the complete financial data of 94 dairy cooperatives for the fiscal year ending in 2007, total assets of these cooperatives were $12 billion (or $8.41 per hundredweight (cwt) of milk, table 1). Current assets accounted for 60.4 percent ($7.3 billion, or $5.08/cwt), and fixed and other assets accounted for the other 39.6 percent ($4.8 billion, or $3.34/cwt). These 94 cooperatives represented 61 percent of all dairy cooperatives and marketed 142.9 billion pounds of milk, or 94 percent of cooperative milk volume.

Total liabilities were $8.7 billion. Of which, 72.3 percent was current liabilities ($6.3 billion or $4.40/cwt), and 27.7 percent ($2.4 billion or $1.69/cwt) was long-term debts. Equities, the balance of assets and liabilities, were $3.3 billion ($2.32/cwt).

Dairy cooperatives typically pay members for their milk twice a month. A large proportion of the current assets and the current liabilities are for such pending periodic cash payments to members. This is a unique characteristic of the balance sheet of dairy cooperatives. Therefore, it is important to focus on the ratio of long-term debts to equity in evaluating financial strength, which was 72.6 percent for the 94 cooperatives.

Equities can be grouped into four categories: common stock, preferred stock, retained earnings, and allocated equities.

Common Stock

In 2007, common stock only accounted for 0.1 percent of total equities (table 1). This is because common stock of cooperatives is usually issued for witnessing membership and carries minimal nominal value.

Preferred Stock

Preferred stock as reported was 7 percent of total equities. A substantial portion of the preferred stock was issued by some cooperatives to members for witnessing retained patronage refunds or for witnessing members' additional investment in the cooperative and may be considered as allocated equities. It is not clear who holds the remaining preferred stock (probably representing less than 5 percent of total equities); the holders could be non-

Table 1. Aggregated balance sheet of 94 dairy cooperatives, 2007

Assets:	$,000	$/cwt	% of category
Current assets	7,258,423	5.08	60.4
Net PPandE and other assets	4,609,394	3.23	38.3
Investments in other cooperatives	152,067	0.11	1.3
Assets not categorized	935	0.00	0.0
Total assets	12,020,819	8.41	100.0
Liabilities and equity:			
Current liabilities	6,290,839	4.40	72.3
Long-term and other liabilities	2,409,129	1.69	27.7
Liabilities not categorized	677	0.00	0.0
Total liabilities	8,699,968	6.09	100.0
Equities			
Common stock	1,857	0.00	0.1
Preferred stock	232,595	0.16	7.0
Retained earnings	358,473	0.25	10.8
Allocated equities	2,727,249	1.91	82.1
Total equities	3,320,174	2.32	100.0
Total liabilities and equities	12,020,142	8.41	
Number of dairy cooperatives reporting	94		
Member milk (million pounds)	142,865		
members as well as members.			

Retained Earnings

Retained earnings could be earnings derived from non-member businesses, but may also include allocated equities that some cooperatives choose not to separately specify in the financial reports, retained net savings that are going to be allocated later, or earnings that are difficult to attribute to specific member transactions. Therefore, retained earnings that are not likely to be subject to allocations (or considered by some to be "permanent" equity) should be less than the reported 10.8 percent of total equities. In any case, retained earnings belong to the cooperative and therefore are owned by members.

In most cases, non-member businesses of dairy cooperatives are incidentals to the dairy operation. These may include:

- Processing into storable products other firms' surplus (distressed) milk that needs to find a home.
- Sales of goods sourced from other firms in dairy stores or other sales outlets.
- Sales of dairy or farm supplies that may include customers who are non-members.

In a limited number of cases, retained earnings are profits from investment activities that may or may not be related to the core business of serving members' marketing and farming needs.

Allocated Equities

The 94 cooperatives reported that 82.1 percent of their equities ($1.91/cwt) were allocated to members. Allocated equities are members' capital from one or more of these sources:

Retained patronage refunds: Retained patronage refunds are net savings that are allocated to members based on patronage but are retained to finance the cooperative's operations after a cash portion has been paid to members. Members must treat the entire patronage refunds (retained as well as cash payment) as income for tax purposes. Cooperatives usually revolve retained patronage back to members after a certain period of time.

Capital retains: Some cooperatives use capital retains to finance the operations or more often, for special projects such as building new plants. Money is withheld from milk payment at a certain rate per hundredweight of milk. Members must treat capital retains as income for tax purposes. Capital retains are also revolved back to members after a certain period of time.

Base capital plan: Some larger diversified dairy cooperatives have adopted base capital plans to establish a more stable equity pool. Under such a plan, a target base capital level is established at a rate per hundredweight of milk marketed during a representative period. The base capital may be funded by retained patronage and/or capital retains, or by other means of member contribution. Once a member attains the prescribed base capital level, future patronage earnings allocated to the member are paid in cash.

Members provide almost all equity capital. Counting common stock, preferred stock (that is issued to members), retained earnings, and allocated equities, almost all equities (probably more than 95 percent) of dairy cooperatives are supplied and owned by members.

SUMMARY—DAIRY COOPERATIVE PRACTICE AND THEORY

Market performance and economic structure of dairy cooperatives are in full accord with the economic theory of cooperation as expounded by Nourse and Emelianoff. Dairy cooperatives' mission, functions, organization, governance, operations, market performance, financing, etc., all conform to the theoretical prescriptions, as table 2 shows. Cooperation as practiced by dairy farmers in marketing milk is an enduring business model that is in full agreement with the economic theory of what cooperatives are and what cooperatives do.

Table 2. Comparison of cooperative theory and dairy cooperative practice

Cooperative Basics I: Market Performance	Market Performance of Dairy Cooperatives
Cooperatives are organized for efficiently carrying out specific business functions.	49,675 dairy farmers in 155 cooperatives marketed 83 percent of U.S. milk in 2007.
Cooperatives can be of any size (and can be local, regional, or national in scope) that allows them to function efficiently in the marketplace.	The smallest local cooperative has a few members marketing less than 1 million pounds of milk per year; the largest one has about 10,000 members in the 48 contiguous States and markets tens of billions of pounds of milk.
Cooperatives afford farmers the organizational sizes for exercising countervailing power.	Dairy cooperatives may grow or have grown to the size necessary for effectively bargaining with milk buyers for better prices and terms of trade.

Table 2. (Continued)

Cooperative Basics I: Market Performance	Market Performance of Dairy Cooperatives
Cooperatives are pro-market; they let the market supply- and-demand price be the guidance for producers.	Dairy cooperatives and their member-farmers are subject to the disciplines of the market in a free economy.
Cooperatives are a means for farmers to promote and maintain competition—as the competitive yardstick.	To be competitive, processors must match the effectiveness and efficiency of dairy cooperatives.
In those fields where the market has become truly competitive and farmers can be well served by other firms, cooperatives may want to cede the field and assume only a stand-by position (to preserve members' capital, time, and efforts for use on the farm), while maintaining the legal institutions and organizational capacity to step in if there is a relapse of market inadequacy.	Dairy cooperatives have comparative advantages in procuring milk and have major shares in making hard products (71 percent of butter, 96 percent of nonfat and skim milk powder, and 26 percent of cheese—the latter decreased from 34 percent in 2002). Their shares are less significant in sectors that are capital-, technology-, and service-intensive and that carry high product and market risks (7 percent of fluid milk, 4 percent of ice cream, 11 percent of yogurt, 14 percent of sour cream. Their share of cheese has also declined in recent years). However, dairy cooperatives have the wherewithal to take up the slack if the market fails to perform well.
Cooperative Basics II: Economic Structure	Economic Structure of Dairy Cooperatives
Cooperative organizations represent the aggregates of economic units.	A dairy cooperative is the aggregate of dairy member-farms.
A cooperative is an agency owned and controlled by members through which they conduct their business.	A dairy cooperative is owned, controlled, and used by members as the milk-marketing arm of their dairy farming business.
Each member-farm fully retains its economic individuality and independence.	Member dairy farms are independent economic units, each making its own business decisions.
The board of directors is elected from among member- farmers.	Directors are members; dairy cooperatives may have non-member directors who usually are non-voting advisors and may be mandated by State laws.

Cooperative Basics I: Market Performance	Market Performance of Dairy Cooperatives
Proportionality and service at-cost are two basic principles.	These principles are applied in every facet of operations that relate to member business.
Members provide advances (i.e., equity capital) for financing the cooperative.	Almost all equities are member capital; ownership of a fraction (a portion of preferred stock) is not discernable from the financial statements.
Patronage refunds are returned to members who have been underpaid or overcharged.	Patronage refunds are net savings returned to members.
Dividend on capital, if any, is interest payment for using members' capital.	Dividends, if paid, are usually on preferred stock, and typically at less than 8 percent.
Being an aggregate of member-farms, the cooperative is neither a horizontal integration of its members nor a vertical integration between the cooperative and its members. It is a third mode of organizing coordination.	There may be some degree of coordination among members as they voluntarily and collectively adapt to market situations. However, this is not the same as vertical or horizontal integration.

FINANCING CHALLENGES OF DAIRY COOPERATIVES

Equities of dairy cooperatives are provided by members. Therefore, a cooperative's ability to raise and retain capital is constrained by members' financial resources and their willingness to advance funds to the cooperative. Managing this unique way of equity financing inevitably generates some internal tension between members and the cooperative. However, members are usually supportive of the financing need if the capital requirement is for the cooperative to carry out its milk marketing functions. Resistance to contributing more equity capital tends to occur when the cooperative embarks on ventures that are considered by members to be extraneous or beyond its stipulated core business of marketing members' milk.

Member Loyalty

For an average member-producer delivering 3.1 million pounds of milk a year in 2007, total allocated equity retained by the cooperative amounted to an estimated $59,000 per member (by using the $1.91/cwt rate). Because retained equities also include those yet to be revolved back to retired members and inactive (former) members, equities actually retained for active members should be somewhat lower than this estimated amount. Still, the sum of capital committed by a member to the cooperative is very substantial.

Members must treat retained capital, when allocated, as income for tax purposes and pay taxes out of their own funds. Although the retains are revolved back to members as permitted by the cooperative's earnings, the revolving period is usually at least a few years. (One cooperative is known to have a revolving period of 6.5 years, probably the shortest among all dairy cooperatives.) Therefore, the present value of the retained capital is diminished due to the fact that taxes on them have to be paid upfront and that the revolving funds to be received in the future are discounted.

Members' perceptions and attitudes towards retained equities may vary with their respective membership status—whether they are active members, retired members, or inactive (former) members, even though they all usually receive the revolved equities on the same revolving schedule, which is determined by the board of directors.

Active Members

Active members may realize the necessity to adequately capitalize the cooperative's operations in order to ensure their milk is effectively and efficiently marketed. Still, retained equities compete with capital needs on the farm. It is only natural that members want as little retains and as short a revolving period as possible. Recent USDA data show the magnitude of capital needs on dairy farms. Among all farm operators, dairy farmers are the most heavily in debt because of the type of inputs used and assets owned. At the end of 2007, 67 percent of dairy farms owed debt that was worth, on average, over $226,000 per farm. Dairy operations accounted for only 2.9 percent of all farms but owed 13.3 percent of reported farm-level debt (*Harris, et al.).*

Retired Members

Retired members may be content with receiving retained equities that are revolved on a steady and regular basis; they may consider such payments as something akin to retirement annuities. However, some may express dissatisfaction that no dividend is paid on the retained equities and the cooperative uses their capital free of charge. And if equity revolving becomes erratic, usually due to the cooperative encountering certain financial difficulties, they may become disgruntled.

Inactive (Former) Members

Inactive members may be farmers who have discontinued membership in the cooperative and made other milk marketing arrangements, who have exited from dairy farming and transitioned into other farming enterprises, or who have discontinued farming altogether. Conceivably, they are the least satisfied with equities being retained. They may need capital for use in other endeavors. As their loyalty to the cooperative has waned or becomes nonexistent, they may deem it meaningless to have the retained equity sitting idly (from their perspective) in the cooperative.

Core Businesses

Like any other business, dairy cooperatives require an adequate level of capital to cash-flow business activities. In addition, it may be necessary for the cooperative to own milk-handling equipment and facilities. They also have to maintain a certain level of equities to satisfy the covenants of lending institutions, because in the course of doing business some debt financing is usually necessary.

Own and Operate Milk-Handling Facilities

In addition to bargaining for milk prices, most major dairy cooperatives also perform milk hauling and operate dairy plants. An adequate amount of capital is needed to invest in the necessary equipment and facilities and to finance plant operations and finished-products marketing.

Members generally are receptive to the capital requirements for these activities, because the expenditures are for the functions that address their main concern: that their milk is assured of a market. Milk is moved from member-farms to the market in a timely fashion. Milk supply that is not sold to milk buyers is processed into storable commodity products, such as butter, milk powder, and cheese. Other value-added products also may be processed from the milk.

Value-Added Processing and Marketing

Producing commodity products usually generates low margins. The margins have been further pressured since the early 1980s as the Federal price-support safety net has been lowered. Furthermore, since the beginning of this millennium, the formulas for pricing regulated milk stipulate fixed margins for manufacturing butter, milk powder, cheddar cheese, and dry whey. These fixed margins are difficult to change even in times of rapidly rising input costs.

For these reasons, for more than two decades many dairy cooperatives have gradually shifted away from making commodity products and put more emphases on operations that add value to milk and milk products, such as making specialty or niche varieties of dairy products, aging or further processing cheese, or extracting milk components for use as ingredients in manufacturing food or beverage products. However, making value-added products requires additional investment in equipment, technology, personnel, marketing, operations, etc. Some cooperatives, while otherwise successful in making the transition, encountered cash-flow issues because the demand for capital was more than their members could afford. In such a case, a merger with a cooperative that had a broader membership and financial base usually could rejuvenate the business for the benefit of all members.

Some other cooperatives exited commodity processing altogether to concentrate on bargaining operations—usually when their plants needed modernization; a few invested in plants operated by business partners to ensure a home for members' milk. Still other cooperatives probably did not realize the market had changed, or started the transition too late, or did not have an adequate strategic plan for the transition. They stumbled along, showing erratic financial results and losing members' support in the process. Eventually, they succumbed to the financial stress and filed for bankruptcy, sold off the cooperative, or merged with other cooperatives.

Milk Marketing-Related and Other Member Services

Field services, market information, and other marketing-related services are usually offered to members as a part of the cooperative's milk-marketing efforts and are accounted for as a part of the cost of doing business. Other member services—such as insurance, retirement program, risk management, etc.—are usually self-sustaining programs and only require minimal financial support from the cooperative.

Extraneous Businesses

Dairy farmers usually will support a cooperative's need for financing if the capital is for a venture that would:

- Solidify the market for members' milk, or
- Help market members' milk, or
- Add value to members' milk, and
- Benefit members the most among all available alternatives of investing the capital.

A cooperative could face financing issues if it invests in ventures that members do not consider to be within the realm of these criteria. Two examples are: investment in seemingly related businesses and investment in extended business ventures. As for the dairy export business, members seem to be ambivalent. On the one hand, they certainly have benefitted from growth in dairy exports. On the other hand, the investment required to develop the export market seems to give them pause.

Investment in Seemingly Related Businesses

An example is a dairy cooperative that is considering whether to acquire a dairy business, which has the following attributes:

1) The dairy products it manufactures are outside the cooperative's current product lineup or competence.
2) Its location is outside the cooperative's membership area.
3) It does not, and will not, use the cooperative members' milk (because of the distance)— dairy inputs for making its products are procured locally and are subject to the local supply-and-demand dynamics.
4) Its major market area for the finished products is distant from the cooperative's trade territory. Thus, its synergy with the cooperative's existing business is minimal.
5) Its production operations require proper expertise to supervise, and the cooperative does not possess this expertise in its existing business.
6) Its consumer products require top management's close attention, which could be lacking if the cooperative's management and control system is not properly structured and staffed.

On the surface, the proposed acquisition appears to be an extension of the cooperative's business of marketing milk and processing dairy products. In reality, it is an investment in an unrelated business; its only relation with the cooperative's existing business is that both are in the dairy industry. It can be said that the investment is not going to help solidify the market for members' milk, help market members' milk, or add value to members' milk. Further, the return on investment is uncertain.

The cooperative may be in solid financial condition and have an adequate amount of member capital to properly market its members' milk. However, investing in the proposed venture is likely to overextend the cooperative's financial resources and jeopardize its sound financial foundation. Because members are usually content as long as their milk is picked up and properly marketed, more member capital contribution for the new venture in all likelihood will not be forthcoming. Under the circumstance, if the board and the management insist on taking on the new venture, they soon are likely to find the venture is causing financial stress, or even the eventual demise of the cooperative. (This is the experience of a cooperative that has recently gone bankrupt.)

Investment in Extended Business Ventures

Some dairy cooperatives may market products that are not derived from milk but are necessary for completing the dairy-related product line, in order to enhance the market position of members' products. Others may use their processing plants to package products such as juices, bottled water, and other nondairy drinks to more fully utilize equipment capacity and spread the fixed cost. These products usually are incidental to the cooperatives' main dairy business and therefore do not require a significant amount of additional financial resources to support them.

The extended business venture scenarios most often occur outside the dairy cooperative sector, but the lessons could be enlightening for dairy farmers. A typical case is a cooperative that has a successful consumer brand and wants to leverage the brand beyond its core expertise and business. For example, a fruit cooperative could extend its juice brand into other beverages, or a tree-nut cooperative could extend its brand into other snack foods, etc. The assumption is that consumers would faithfully translate their brand loyalty into buying the new categories of products.

However, investment in the extension of the brand may or may not be beneficial to cooperative members. It depends on whether the investment would:

- help solidify the market for members' products, or
- help market members' products — depending on how members' products are incorporated in the product mix, or

- add value to members' products — depending on how the extended categories of products are formulated, and
- benefit members the most among all available alternatives of investing the capital.

Furthermore, competing in the consumer market beyond marketing members' products would require ample additional capital for market research, product development, process design, sales and promotion, and marketing logistics, etc. Often, members' equity is not sufficient to support the extended business venture.

Some cooperatives try to find additional equity capital from outside investors. Other cooperatives could convert into investor-owned firms (most likely as publicly traded companies). Either way, converting the composition of equity capital sources would also change the character and the priority of the organization. In planning such conversions, members should consider whether in the pursuit of profit to placate investors, their production will still be adequately priced and marketed. Members of dairy cooperatives in the United States have not pursued conversion to investor ownership, as have some cooperatives in other commodity sectors. They expect dairy cooperatives to market their milk effectively and efficiently, and little more.

Investment in Developing Dairy Export Markets

The United States has not been a major exporter of dairy products on a sustained basis. However, in 2007- 08, due to tighter global stocks, drought-induced production declines in Oceania, rising demand in foreign countries, and the weaker dollar, the United States was able to take advantage of significant export opportunities (*USDA/ERS; Ling, 2008*). Commercial exports, on a milk-fat basis, more than doubled from 3.4 billion pounds of milk equivalent in 2006 to 8.7 billion pounds in 2008, or an increase of 5.3 billion pounds in 2 years (table 3). On a skim-solids basis, exports increased by 3 billion pounds of milk equivalent, from 23.6 billion pounds in 2006 to 26.6 billion pounds in 2008.

Export demand raised milk price to an unprecedented high level for an extended period from spring 2007 through fall 2008. The all-milk price reached its record high of $21.90 per cwt in December 2007. This provided incentives for milk production to expand. Meanwhile, the factors that were favorable to U.S. dairy exports lapsed. When the worldwide recession hit in late 2008, milk price plummeted. The all-milk price dropped to $11.30 per cwt in June and July 2009, a decline of more than $10 per cwt from the 2007 peak.

Table 3. U.S. commercial exports of dairy products, 2004-2009

	2004	2005	2006	2007	2008	2009
	Billion pounds					
Milk equivalent, fat basis						
Milk production	170.9	176.9	181.8	185.7	190.0	189.3
Commercial exports	3.4	3.3	3.4	5.7	8.7	4.5
Export share	2.0%	1.9%	1.9%	3.1%	4.6%	2.4%
Milk equivalent, skim-solids basis						
Milk production	170.9	176.9	181.8	185.7	190.0	189.3
Commercial exports	16.0	19.3	23.6	24.5	26.6	22.4
Export share	9.4%	10.9%	13.0%	13.2%	14.0%	11.8%

Source: USDA World Agricultural Supply and Demand Estimates, November 2010 and various previous issues.

Milk price has since recovered, but price volatility can be expected to continue because of the ever-shifting supply-and-demand situation. Drawing from the experience of high milk prices during the 2007- 2008 dairy export boom, some have suggested that to maintain sustained prosperity, the U.S. dairy industry should strive to be a consistent supplier in the export market (*e.g., Innovation Center for U.S. Dairy*).

International dairy trade absorbs about 5 percent of milk produced globally. The trade is primarily in major manufactured dairy products—butter, cheese, and dry milk powders—with some trade in fluid milk products, ice cream, yogurt, and dry whey products (*USDA/ERS*). Although the trade volume of dry whey products is significant, its value is relatively low.

There are basically two approaches to international trade (*Field*):

1) Trading products in the international dairy market. The bulk of the trade is in commodities such as butter, cheese, and dry milk powders.
2) Being a direct participant in the market of a target country. Some multinational firms may have investment in dairy processing facilities and even in dairy farms in the target country to produce products for the local market. Some of the dairy ingredients used in local processing may be imported from other countries by the multinational firm.

Most U.S. dairy exports are in bulk commodities. Because the domestic market is vast, and historically the domestic price level has been high relative to the international prices, the dairy industry tends to regard the export market as the last-resort outlet, while the world treats the United States as a residual supplier.

Even the spike in exports in 2007-2008 did not change this basic relationship, although such recent export experience may gradually transform U.S. dairy exporters into more consistent suppliers.

Some of the factors that are required of a consistent exporter are (*Field*):

- In-country contacts and market understanding.
- Competitive pricing.
- Appropriate product range.
- Strong partners (distribution, technology).
- Proactive approach to business development, sales, and marketing.
- Scale and focus of business.

In other words, to be a consistent exporter would require considerable effort and financial resources to develop and maintain the export market.

To have members' support, the challenge is to convince them that investing in the undertaking would help solidify the market for members' milk, help market members' milk, or add value to members' milk, and benefit members the most among all available alternatives of investing the capital.

As for becoming a direct participant in the market of a target country, it probably will not happen soon. Although dairy exports have recently shown encouraging growth, the share of U.S. domestic consumption is still overwhelming.

It would be difficult to convince members that investing in dairy plants and farms abroad is a judicious use of members' equity, unless it could be shown that such investment is necessary for the purpose of protecting and expanding the market for members' milk and milk products.

Equity Financing Alternatives

There is no doubt that equity capital is often the constraining factor in the ability of a cooperative to make new investment. Some cooperatives have tried alternative equity financing methods to leverage cooperative members' capital. Examples are forming a public stock corporation, issuing preferred stock, or forming a limited liability company, joint venture, or "new-generation" cooperative, etc.

Public Stock Corporation

There is one known case of a dairy cooperative offering common stock in one of its subsidiaries in the late 1980s. The cooperative converted its fluid business subsidiary into a publicly traded stock company, the idea being to use investor financing and stock as tools for expansion and growth, while members maintained the majority ownership of the business. However, in less than 3 years, the cooperative bought back all outstanding stock from minority owners.

It can be difficult for a cooperative to operate a public stock corporation subsidiary because there are fundamental conflicts between benefits for member-producers and investors' focus on returns on investment. In the dairy business, the conflict between producer milk pay price and profit for investors may be difficult to reconcile. Furthermore, with investor capital, the subsidiary and even the cooperative may lose Capper-Volstead status in inter-state commerce.

Preferred Stock

A cooperative may issue preferred stock to raise more funds from members or to tap nonmember capital. Preferred stocks that pay dividends and have preference in assets over common stock in the event of the dissolution of the cooperative are the most common type. Some preferred stocks may be considered as equity capital, while others may look more like debt capital, depending on how the rights of the shareholders are specified.

Limited Liability Company (LLC)

An LLC is a State-approved, unincorporated association, just like a partnership except that it protects its owners and agents from personal liability for debts and other obligations of the LLC. Earnings pass through to the owners (no non-qualified retains) and enjoy single-tax treatment. An LLC may operate on a cooperative basis. Or it may allocate earnings and losses and assign votes among its owners any way they want to. Some producers believe that an LLC provides greater flexibility for tapping investor capital. However, the combination of producers and investors in an LLC would encounter the same issues as in a publicly traded subsidiary operated by a cooperative.

Joint Venture

An LLC may be a useful model for established cooperatives to form joint ventures with other cooperatives or firms. On the marketing side, a joint venture LLC may be used by a cooperative and its partner to develop and market certain dairy products. The cooperative supplies dairy inputs and the partner provides technical or marketing know-how to the LLC. The joint-venture partners share the financing and the risk of the business activities of the LLC. This organizational model reduces the cooperative's capital requirement and risk exposure, while a market outlet for milk is secured. Many recent joint ventures formed by cooperatives with other business entities are organized as LLCs.

"New-Generation" Cooperative

A new-generation cooperative usually requires significant equity investment as a prerequisite to membership and delivery rights, in order to ensure that an adequate level of capital is raised and the plant capacity is fully utilized. The delivery right is in the form of equity shares that can be sold to other eligible producers at prices agreed to by the buyer and seller, subject to the approval of the board of directors. The transferable delivery right is appealing to members because it allows them to cash in on any increase in the value of their cooperative when they retire. Interest in new-generation cooperatives surged in the 1980s and 1990s, largely in response to the market condition prevailing during that time period. Cooperative development leaders believed that this form of cooperative organization would solve the problem of depressed farm income by engaging in value-added processing. Many new-generation cooperatives have been successful. However, the attributes of the new-generation cooperative model also have created some problems, mainly related to delivery-right and property-right issues (*Torgerson*). After the turn of the 21st century, interest cooled down substantially. There was only one dairy cooperative known to have been organized using the new-generation model. In 1995, Dakota Dairy Specialties was established to make specialty cheese. But its remote location, the capital investment needed to renovate its plant, and the skill required to make and market specialty cheese posed major problems, and the new-generation model proved no help. It suffered the same fate as the struggling cooperative it was formed to replace. By 1999, Dakota Dairy Specialties ceased to operate.

ADDRESSING COOPERATIVE EQUITY FINANCING ISSUES

Dairy cooperatives may be characterized as epitomizing the economic theory of what cooperatives are and what cooperatives do (the classic *raison d'être* of cooperatives). Their experience as shown in this report can serve as an example for addressing equity financing issues that were raised in the introduction of this report.

1) ***Question.*** Because equity capital is furnished by members and therefore is rather limited, how does a cooperative gain access to capital without incurring long-term debt to fund the organization in this new, expanded and globalized marketplace?

Answer. Members organize or join a cooperative to market their farm production. It is in their interest to provide capital for the cooperative's operation. But they need to be convinced that the cooperative's venture is necessary to solidify the market for their products, sell more of their products, or add value to their products. They must also agree that the benefit of the endeavor is the best alternative for investing the capital.

Members of a newly formed cooperative may need some assistance in the initial stage of its formation. Over time, though, the cooperative must be self-sustaining.

In both cases, the information about market reality and a well thought-out business plan that the cooperative intends to follow should be made clear to the membership. After weighing all available options, it is up to members to decide if they want a viable cooperative to market their products and support it with an adequate level of equity capital, or if they prefer other alternatives or business models.

2) ***Q.*** How does a cooperative provide a vehicle for members to gain access to the value of the business, which some of them perceive to be higher than its book value, without selling the cooperative or going public?

A. If market value of the cooperative is higher than the book value, it means the cooperative's earnings and potential future earnings are higher than can be expected, given its level of equity capital. This usually reflects certain attributes the cooperative possesses, such as: unique and highly desirable products; profitable market niches; valuable intellectual properties (technology, manufacturing and marketing know-how, brand names, etc.); and superb governance, management and staff. Members gain access to the cooperative's value of higher earning ability by receiving higher pay prices, premiums, and patronage refunds. Selling the cooperative to gain the value of the business is tantamount to "killing the goose that lays the golden egg."

The agitation for accessing the perceived market value of the cooperative most often comes from members who are near retirement or whose farming business is otherwise approaching the end of the line. However, some active members may also prefer the instant payout. Other members may want the cooperative to continue marketing farm production for the current generation and the generations to come. Basically, it is up to members to decide whether retaining the cooperative or selling it is in their best interests.

3) ***Q.*** Facing equity redemption issues or in need of more capital, are there alternatives short of selling the cooperative or going public?

A. Members have substantial equity invested in the cooperative for marketing their products effectively and efficiently. Equity investment should be commensurate with the functions they want the cooperative to perform on their behalf. Constrained by insufficient equity capital or having difficulties raising more capital, the cooperative may have to retrench.

Through the years, many dairy cooperatives exited from manufacturing and processing dairy products when their plants needed modernization and the required milk volume and capital investment for the new plant "outsized" the cooperative. They transformed the cooperative to focus on bargaining operations to fulfill members' primary goal that their milk be assured a market and receive a fair pay price. Many other cooperatives merged to have a larger pool of milk volume and capital, which could be deployed more efficiently as a result of economies of scale.

Bargaining cooperatives usually require less capital to operate. In 2007, bargaining-only cooperative members had just $0.42 of equity per cwt of milk marketed through their cooperatives, while members of niche-marketing cooperatives had $4.78 per cwt and members of diversified and fluid processing cooperatives had $2.79 (*Liebrand*).

It is up to members to decide what they want the cooperative to do and how much they want to support it, or if they prefer other alternatives or business models.

4) ***Q.*** Are new-generation cooperatives the answer to cooperative financing issues?

A. The primary difference between a new-generation cooperative and a traditional one is that members of a new-generation cooperative have to pay the required equity upfront to acquire delivery rights; a traditional cooperative accumulates equity over time through retained patronage. While this and other attributes have helped many new-generation cooperatives achieve successes, they also encountered their own set of issues, mainly relating to delivery rights and property rights.

One further issue that has not been discussed is the venture-capital character of the investment in new-generation cooperatives. Many new-generation cooperatives are organized to take advantage of an investment opportunity that promises enticing returns by processing one product or a narrow range of products. The venture is capital intensive and requires a large start-up fund. The expected returns on investment may be high, but so may be the risk.

An example is a cooperative organized by corn producers to invest in an ethanol plant that will use members' corn as feedstock. In this case, members' equity in the cooperative is very much like venture capital and members' corn is tied to a single use. Both of these attributes entail substantial risks, including risks that are related to unsettled ethanol technology, uncertain ethanol market outlook, volatile corn-ethanol-petroleum market dynamics, and shifting priorities of public policies (subsidies, tariffs, etc.). This mode of operation is very different from a grain cooperative that is organized to market members' corn through a variety of marketing channels, perhaps with supplying ethanol plant(s) as one of its enterprises.

So, new-generation cooperatives have their pros and cons but probably are not a panacea for cooperative financing issues. Furthermore, consideration should be given to whether a cooperative (new-generation or otherwise) is most appropriate for organizing a particular new venture or if some other business model is more suitable.

5) ***Q.*** What effects might issuing preferred stock have on a cooperative's practice?

A. Preferred stock may specify nearly any conceivable right for shareholders. What effects preferred stock may have on a cooperative's practice depend on what rights are specified. Preferred stock that pays dividends and has preference in assets over common stock in the event of the dissolution of the cooperative— the most common type of preferred stock— probably would not have any impact. If the preferred stock confers certain voting rights, the effect would depend on what specific issues the preferred stock holders are entitled to vote on.

6) ***Q.*** What long-term effects can (unallocated) retained earnings from non-member business have on cooperative governance?

A. Cooperatives may have non-member business for various reasons. In any case, retained earnings sourced from non-member business are owned by the cooperative and, therefore, jointly owned by members. Disposition of retained earnings is at the discretion of the board of directors. However, a cooperative would not be conforming to the Capper-Volstead Act requirement if its nonmember business exceeds 50 percent of total sales.

In the case of dairy cooperatives, retained earnings represented 10.8 percent of member equities in 2007 (table 1). However, besides earnings that are sourced from non-member business, a substantial portion of these earnings may include: allocated equities that some cooperatives choose not to separately specify in the financial reports; retained net savings that are going to be allocated later; or earnings that are difficult to attribute to specific member transactions.

7) *Q.* What changes in governance, organizational structure, and practice may be brought about by the new cooperative laws enacted by some States that allow outside (non-member) equity capital?

A. There is a large variation among the few State laws that allow cooperatives to have outside investors. They vary from reserving the voting power to member-patrons only, to setting a minimum level of voting power for member-patrons. Requirements regarding earning distribution between member-patrons and investors also differ substantially. Differences in governance and earning distribution rules and the type of investors involved (e.g., for-profit investors, non-profit economic development organizations, etc.) will have different influences on cooperative organizational structure and cooperative practice. It is probably better to analyze them on a case-by-case basis.

Not every cooperative newly incorporated under these State laws has investors; some choose not to have investors and operate as a traditional cooperative.

8) *Q.* How plausible is the contention that there is a large amount of untapped equity in rural America that cooperatives do not have access to and that should be allowed to be invested in cooperatives?

A. If the farm sector equity is any indication, it is not clear how much of it is untapped or available for off-farm investment. The equityto-asset ratio of the farm sector is between 88 and 90 percent during the 5-year period 2006- 2010 (table 4). This is in contrast to around 96 percent of the total assets that are in fixed and farming assets (about 84 to 85 percent of the assets are in land and real estate and 11 to 12 percent are in farming assets). Only around 4 percent of the assets ($74 billion to $85 billion) are in financial assets, and they may not be available for investing in new off-farm ventures unless the expected return from the new investment could out-perform the opportunity cost of on-farm working capital requirement or the opportunity return of existing financial investments. Borrowing against the assets to invest in new off-farm ventures may not be a promising proposition.

Of course, the farm sector does not represent the entire rural economy, but comparable financial data for rural America is not readily available. Recent assessment of the state of rural economy shows that it faces significant challenges (*Council of Economic Advisers*). In any case, it is highly unlikely that there is a large amount of untapped equity to be found in rural areas.

CONCLUSION

Dairy cooperatives are prime examples of the traditional model of a cooperative that is owned, controlled, financed, and used by members. Focusing on the business of marketing members' milk, dairy cooperatives benefit members by enhancing returns to their milk production efforts; members supply equity capital needed for the cooperative to carry out its function as their collective milk marketing arm.

The cohesiveness between member purposes and cooperative functions makes dairy cooperatives, as a group, perhaps the most prominent agricultural marketing cooperatives.

This is because milk is highly perishable and its daily production must have an assured, ready market.

To most dairy farmers (84 percent of U.S. total in 2007), marketing services provided by their cooperatives are indispensible for the dairy farming business.

It is for this reason that equity capital financing, in general, is not a contentious issue for dairy cooperatives if the fund is for the core business of marketing members' milk.

This case study shows that dairy cooperatives are seldom used as a vehicle for investing in ventures that are unrelated to member business.

The close bond between producers and their dairy cooperatives may or may not be replicated in other agricultural commodity sectors, depending on the characteristics of the commodity and its market. Because no two commodities are the same, the needs of respective producers in marketing them also vary. Cooperatives may be more essential to producers of commodities that have to be marketed shortly after being produced (such as vegetables, fruits and, of course, milk) or that have no ready market outlet other than the cooperative, than they are to producers of commodities that are storable and have a longer marketing season (such as grains and oil seeds) or that have multiple market outlets.

Table 4. U.S. farm sector balance sheet selected items, 2006-2010[F]

	2006	2007	2008	2009[P]	2010[F]
			$ billion		
Farm assets	1,924	2,055	2,016	2,044	2,096
Real estate	1,626	1,751	1,703	1,727	1,777
Financial assets	74	79	82	84	85
Farming assets[1]	224	225	231	232	233
Farm debt	204	214	243	245	245
Farm equity	1,720	1,841	1,773	1,798	1,861
Selected ratios:			*Percent*		
Real estate/assets	84.5	85.2	84.4	84.5	84.8
Financial assets/assets	3.8	3.8	4.0	4.1	4.0
Farming assets[1]/assets	11.7	11.0	11.5	11.4	11.1
Equity/assets	89.4	89.6	88.0	88.0	88.8
Debt/equity	11.8	11.6	13.7	13.6	12.6
Debt/assets	10.6	10.4	12.0	12.0	11.2

F = forecast and P = preliminary. [1]Livestock and poultry, machinery and motor vehicles, crops stored, and purchased inputs.

Source: *Farm Income and Costs: Farm Sector Income Forecast*, Economic Research Service Brief Rooms, updated August 31, 2010.

It stands to reason that raising or retaining equity capital is more challenging for a cooperative that is regarded by its members as but one of the competing market outlets for their products than for a cooperative that is indispensible to members.

It can be even more challenging for a farm supply cooperative that has to compete with other supply stores in the local market. There are hundreds, or even thousands, of supply items, and it is unlikely that the cooperative can be the best-value provider of every piece of merchandise. “Cherry-picking” by members in making purchases is inevitable. However, it is difficult to raise equity capital from members in this circumstance. (A food cooperative competing with other stores may encounter the same issue.)

Regional farm supply cooperatives may have economies of scale in product sourcing or in operating manufacturing facilities, especially for major supply items such as seeds, feed, fertilizer, chemicals, and petroleum products.

They could pass along cost-savings derived from scale economies to members and thus better meet competition. However, operating upstream manufacturing plants has its own risks (such as volatile raw material prices) that require the cooperative to have ample capital to cushion the shocks. The challenge for these cooperatives is to have a solid and broad membership base that sees the value of supporting the cooperatives with adequate equity capital.

All these point to the fact that the cooperative capital financing issue is really a reflection of a certain gap or disconnect between member purposes and cooperative functions. It is less of an issue the narrower the gap and a more serious issue the wider the gap. The solution to the issue then lies in assessing what members want the cooperative to do and whether they are willing to finance it with equity capital in the amount commensurate with the benefits they expect to receive—and the cooperative should operate accordingly for members' best interests. In some cases, they may decide whether the cooperative is the most suitable business model for what they want to accomplish.

In recent years, the cooperative model has gained new attention from social entrepreneurs and economic development practitioners. Being owned, controlled, and used by members for mutual benefits, cooperatives are an appealing tool to empower people to work toward their own economic destiny. They can be adapted to be community-based organizations to serve economic opportunity-deprived or service-deprived areas. Because such cooperative organizations are formed to address public policy or social issues, it is appropriate to have initial capital funding assistance from public or philanthropic sources. Over the long term, however, they must be self-sustainable in order to be economically viable. Some exemplary precedents are rural electric cooperatives (*National Rural Electric Cooperative Association*) and the Farm Credit System (*Farm Credit Administration*).

REFERENCES

American Heritage Dictionary of the English Language, New College Edition, Houghton Mifflin Company, Boston, 1976.

Council of Economic Advisers, Executive Office of the President. *Strengthening the Rural Economy*, April 2010.

DeVille, Katherine C., Jacqueline E. Penn, and E. Eldon Eversull. *Cooperative Statistics, 2008*, U.S. Department of Agriculture, Rural Development, Service Report 69, November 2009.

Emelianoff, Ivan V. *Economic Theory of Cooperation: Economic Structure of Cooperative Organizations*, Washington, D.C. 1942 (litho-printed by Edwards Brothers., Inc., Ann Arbor, Michigan), 269 pages. (A reprint by the Center for Cooperatives, University of California, 1995, may be accessed at: http://cooperatives.ucdavis.edu/reports/index.htm.)

Farm Credit Administration. *History of the FCA and the FCS, http://www.fca.gov/about/history/historyFCA_FCS.html.*

Field, Richard. "Can the U.S. Be a World Market Player?" *Hoard's Dairyman*, March 10, 2009, p. 173.

Galbraith, John Kenneth. *American Capitalism: The Concept of Countervailing Power*, revised edition, Houghton Mifflin Company, Boston, 1956; esp. Chapter XI, The Case of Agriculture, pp. 154-165.

Harris, J. Michael, James Johnson, John Dillard, Robert Williams, and Robert Dubman. *The Debt Finance Landscape for U.S. Farming and Farm Businesses*, AIS 87, USDA, Economic Research Service, November 2009.

Hess, Jerry N. *Oral History Interview with Dr. Edwin G. Nourse*, Washington, D. C., March 7, 1972, Harry S. Truman Library, Independence, Missouri. http://www.trumanlibrary.org/oralhist/nourseeg.ht m.

Innovation Center for U.S. Dairy. *The Impact of Globalization on the U.S. Dairy Industry: Threats, Opportunities, and Implications*, Arlington, VA., October 2009.

Larsen, Grace H. and Henry E. Erdman. "Aaron Sapiro: Genius of Farm Co-operative Promotion," *The Mississippi Valley Historical Review*, Vol. 49, No. 2, 1962, pp. 242-268.

Liebrand, Carolyn B. *Financial Profile of Dairy Cooperatives, 2007*, USDA Rural Development, Research Report 219, April 2010, table 4.

Ling, K. Charles. *Marketing Operations of Dairy Cooperatives, 2007*, USDA Rural Development, Research Report 218, July 2009.

Ling, K. Charles. "High prices for dry dairy products due to diverse reasons," in *Whey to Ethanol: Biofuel Role for Dairy Cooperatives?* USDA Rural Development, Research Report 214, February 2008, p. 4.

National Rural Electric Cooperative Association. *History of Electric Co-ops*, http://www.nreca.org/AboutUs/Coop101/CoopHistory.htm.

Nourse, Edwin G. "The Economic Philosophy of Co-Operation," *American Economic Review*, Volume XII, No. 4, December 1922, pp. 577-597.

Nourse, Edwin G. "The Place of the Cooperative in Our National Economy," *American Cooperation 1942 to 1945*, American Institute of Cooperation, 1945, pp. 33- 39.

Sapiro, Aaron. "True Farmer Cooperation," *Journal of Agricultural Cooperation*, Volume 8, 1993, pp.81-93, reprinted from the *World's Work*, May 1923, pp. 84-96.

Shaffer, James D. "Thinking about Farmers' Cooperatives, Contracts, and Economic Coordination," *Cooperative Theory: New Approaches*, ACS Service Report Number 18, U.S. Department of Agriculture, Agricultural Cooperative Service. July 1987, pp. 61-86.

Torgerson, Randall E. "A Critical Look at New-Generation Cooperatives," *Rural Cooperatives*, United States Department of Agriculture, January/February 2001, pp.15-19.

U.S. Department of Agriculture. Economic Research Service Brief Rooms, Dairy: Trade, http://www.ers.usda.gov/briefing/Dairy/Trade.htm.

U.S. Department of Agriculture. Rural Development, Business and Cooperative Programs Public Meeting on Research Transcript: September 27, 2005: http://www.rurdev.usda.gov/rbs/pub/ResearchPubli cMeeting Transcript.pdf.

In: Dairy Cooperatives Profiles and Research ISBN 978-1-62081-247-1
Editors: A. West and S. I. Kelly

Chapter 2

MARKETING OPERATIONS OF DAIRY COOPERATIVES, 2007*

K. Charles Ling

ABSTRACT

A total of 49,675 member-producers of the Nation's 155 dairy cooperatives marketed 152.5 billion pounds of milk, or 82.6 percent of all milk marketed, in 2007. Forty-five cooperatives operated 176 dairy processing and manufacturing plants, 12 handled milk through receiving stations only, and 98 had no milk handling facilities. Cooperatives marketed 71 percent of the Nation's butter, 96 percent of nonfat and skim milk powders, 26 percent of natural cheese, 7 percent of packaged fluid milk products, 4 percent of ice cream, 13 percent of ice cream mix, 11 percent of yogurt, 42 percent of dry whey products, 14 percent of sour cream, and 20 percent of condensed buttermilk. Cooperatives in the 1-billion-to-2-billion-pounds of milk group showed the most significant increase in the share of total cooperative milk volume, while the milk share of the larger sized groups, as a whole, was slightly lower. A total of 86 dairy cooperatives reported having 21,475 full-time and 2,944 part-time employees. Complete financial data submitted by 94 dairy cooperatives showed that total assets for the fiscal year were $8.41 per hundredweight (cwt); total liabilities, $6.09 per cwt; members' equity,

* This is an edited, reformatted and augmented version of USDA Rural Development, Research Report 218 July 2009, dated July 2009.

$2.32 per cwt; and net margin before taxes, 28 cents per cwt, which represented a return on equity of 12.2 percent.

PREFACE

Information for this report came primarily from the responses to a once-every-5-year USDA survey of all U.S. dairy cooperatives. The dairy-specific data were collected in conjunction with the annual survey of all cooperatives by the Cooperative Programs' Statistical Services staff of USDA Rural Development. In some cases, data were estimated for non-respondents based on their financial statements or other sources.

Cooperatives were asked to supply information on their milk-marketing operations for the fiscal year ending in calendar 2007. These fiscal years vary within the calendar year, so the data reflect some differences in time periods.

This report also updates and revises Rural Business-Cooperative Service (RBS) Research Report 201, which was based on cooperative operations for the fiscal year ending in calendar 2002. Revisions were made for consistency in interpreting data between the two surveys.

Unless otherwise specified, when calculating cooperative marketing shares of various products, information for U.S. total volumes was from the statistics reported by USDA National Agricultural Statistics Service: *Milk Disposition and Income, Final Estimates 1998-2002, May 2004; Milk Disposition and Income, Final Estimates 2003-2007, May 2009; Dairy Products, 2003 Summary, April 2004; and Dairy Products, 2008 Summary, May 2009.*

The cooperation of the responding cooperatives and other persons who supplied the necessary data for this report is gratefully acknowledged.

The historical summaries of the results of this and previous surveys may be accessed at http://www.rurdev.usda.gov/rbs/coops/dairy.htm.

HIGHLIGHTS

Member-producers of dairy cooperatives marketed 152.5 billion pounds of milk in 2007, a 9.6 percent increase from 2002. This volume represented 82.6 percent of the milk marketed by farmers nationally, a slight increase from 82.4 percent 5 years earlier. The number of dairy cooperatives during this period decreased from 194 to 155. There were 45 cooperatives that processed and manufactured dairy products, the same number as in 2002, while 12 cooperatives operated receiving stations only and 98 had no milk-handling facilities.

Sixty-three percent of total cooperative volume was sold as raw milk in 2007 versus 61 percent in 2002. The other 37 percent was manufactured at plants owned and operated by cooperatives.

There were 49,675 member producers marketing milk in 2007, 19 percent (11,715) fewer than 5 years earlier. Three regions - East North Central, North Atlantic, and West North Central - together accounted for 85 percent of all member producers and 51 percent of cooperative milk volume. The Western region was the top source of cooperative milk. At 58.1 billion pounds, it represented 38 percent of all cooperative milk.

Dairy cooperatives owned and operated 193 plants, 17 for receiving and shipping milk only, 34 for manufacturing American cheese, 17 for making Italian cheese, 49 for packaging fluid milk products, 24 for churning butter, 39 plants for drying milk products, and 24 for whey products. Other plants made various other dairy products. (A plant may perform more than one function.)

Volumes of butter and nonfat and skim milk powders increased from 2002 to 2007. Cooperatives' share of butter, at 1,087 million pounds, remained at 71 percent of U.S. production, and their share of nonfat and skim milk powders, at 1,444 million pounds, was an overwhelming 96 percent. However, cheese made by cooperatives had a substantial drop, decreasing by 15 percent from 5 years earlier to 2,513 million pounds and accounting for 26 percent (versus 34 percent in 2002) of U.S. production. Cooperatives' share of dry whey products also declined, from 52 percent to 42 percent in 2007.

Sales of packaged fluid milk products by cooperatives increased both in volume and in market share. The 4,035 million pounds marketed was 7.4 percent of the Nation's production, up from 7 percent in 2002. Share of ice cream increased from 3 percent to 4 percent, while share of ice cream mix increased from 6 percent to 13 percent. In 2007, cooperatives marketed 11 percent of the Nation's yogurt, 14 percent of the sour cream, and 20 percent of the condensed buttermilk. As in 2002, there were four cooperatives that each handled more than 6 billion pounds of member milk in 2007. The four accounted for 49.2 percent of cooperative milk volume in 2007, the same share as reported for 2002. Together, the 17 cooperatives in the more-than-2-billion-pounds-size groups had a very slight decrease in the share of cooperative milk, 0.3 percentage point, from 80.6 percent in 2002 to 80.3 percent in 2007. Both the cooperative number and the milk volume of the 1-billion-to-2-billion-pounds group more than doubled over the 5-year period. This is the size group to show the most significant increase in the share of total cooperative milk volume (from 5.1 percent in 2002 to 10.1 percent), mostly at the expense of the groups of cooperatives with smaller milk volumes. The four largest cooperatives had only a slightly higher share of the Nation's milk, moving from 40.5 percent in 2002 to 40.7 percent in 2007. Broadening the focus to the largest 8 dairy cooperatives and the largest 20, their shares of cooperative milk or total U.S. milk changed little or not at all. In terms of milk volume, the relative position of dairy cooperatives to the rest of the industry has been remarkably stable. Sixty-five dairy cooperatives reported having a total of 21,475 full-time and 2,938 part-time employees in 2007. Six other cooperatives each had only one part-time employee. Another 15 cooperatives reportedly had no employees. These 86 cooperatives marketed 86 percent of cooperative milk. Complete financial data submitted by 94 dairy cooperatives showed that total assets for the fiscal year ending in 2007 were $12 billion ($8.41 per cwt), total liabilities were $8.7 billion ($6.09 per cwt), and members' equity was $3.3 billion ($2.32 per cwt). Eighty-two percent of the equity was allocated to members. Net margin before taxes was $404 million (28 cents per cwt), a return on equity of 12.2 percent. Together these cooperatives marketed 94 percent of total cooperative milk volume.

INTRODUCTION

Farmer owned and operated dairy cooperatives continue to provide the most significant channel for marketing milk from the Nation's dairy farms. In line with industry and dairy farm trends, the number of cooperatives is declining, but those remaining are handling larger volumes. In addition, many cooperatives own and operate plants to process, manufacture, and market various dairy products. This report, ninth in a series of periodic appraisals of the scope and performance of dairy cooperatives, describes their continuing evolution in an ever-changing market environment.

COOPERATIVE INDUSTRY PROFILE

Between 2002 and 2007, the number of dairy cooperatives decreased by 39 (20 percent), from 194 to 155. The pace was faster than the reduction of 32 cooperatives (14 percent) recorded from 1997 to 2002. In 2007, 129 (83 percent of all dairy cooperatives) were headquartered in the North Atlantic, East North Central, and West North Central regions; the North Atlantic region had the most with 55 dairy cooperatives (table 1). These three regions also showed a drop of dairy cooperative numbers from 2002. The steepest reduction was 30 in the North Atlantic region. The South Atlantic region had an increase of two cooperatives. The number of dairy cooperatives in the South Central and Western regions remained unchanged.

In 2007, 45 cooperatives (29 percent of all dairy cooperatives) processed and manufactured dairy pro ducts, the same number as in 2002. The number of cooperatives with plant operations increased by one both in the South Atlantic and South Central regions, and it decreased by one each in the North Atlantic and West North Central regions. The East North Central region still had the most cooperatives (20) with plant operations.

Cooperatives that operated only milk receiving stations decreased from 19 to 12 by 2007. The West North Central region dropped five cooperatives, while the East North Central and South Central dropped two each. There was a gain of two cooperatives that operated only milk-receiving stations in the Western region.

Cooperatives that did not have plants or milk-receiving facilities decreased from 130 to 98. Almost half of these were in the North Atlantic region. However, the 48 such cooperatives in that region re presented a decrease of 29 from 2002, the steepest decline among the 3 regions that showed a decline. The South Atlantic, West North Central, and South Central regions each gained one such cooperative.

Most cooperatives (88 percent of all dairy cooperatives) sold at least some bulk raw milk (table 2). The number of cooperatives that marketed butter, Italian cheese, sour cream and yogurt remained unchanged from 2002.

There was a gain of four cooperatives that made specialty cheeses and a gain of one each that made dry whole milk and condensed milk. On the other hand, six fewer cooperatives than in 2002 made American cheese. Cooperatives that handled other selected dairy products all declined in number.

MILK VOLUME AND UTILIZATION

In 2007, cooperative member- producers marketed 152.5 billion pounds of milk, or 82.6 percent of milk marketed by all U.S. producers (table 3). The cooperative share of milk was an increase of 0.2 percentage point from 2002, when it was 82.4 percent. Over the 5- year period, milk marketed by cooperative members increased 9.6 percent, while the increase was 9.2 percent by all U.S. producers.

Milk received from nonmembers and non-cooperative firms decreased by almost 2 billion pounds (37 percent) when compared with 2002. In total, cooperatives received or bargained for 155.8 billion pounds of milk, or 84.4 percent of total volume marketed by all U.S. producers. Cooperatives' share was 1 percentage point lower than in 2002.

Cooperatives sold 63 percent of this volume as raw milk, up 2 percentage points from 2002 (table 4). In other words, cooperatives processed or manufactured 37 percent of the milk in the plants they directly operated.

Market strategy changes and industry restructuring in recent years caused a large volume of the raw milk to be shipped to dairy plants in which cooperatives had an investment but did not directly operate.

To some extent, declining cooperative plant numbers and decreases in market shares of some products were the results of this market evolution.

Figure 1. Number of Cooperatives Operating in Each Region, Member-Producers, Member Milk, and Milk per Producer, 2007.

Table 1. Dairy cooperatives by type of operation and by headquarters region, 2002 and 2007

Region[1]	Processing and manufacturing dairy products		Operating milk receiving facilities only		Not operating milk plants or receiving facilities		Total	
	2002	2007	2002	2007	2002	2007	2002	2007
				Number				
North Atlantic	7	6	1	1	77	48	85	55
South Atlantic	1	2	0	0	2	3	3	5
East North Central	20	20	5	3	16	12	41	35
West North Central	10	9	10	5	24	25	44	39
South Central	0	1	2	0	1	2	3	3
West	7	7	1	3	10	8	18	18
All regions	45	45	19	12	130	98	194	155
				Percent				
Share of total cooperatives	23	29	10	8	67	63	100	100

[1]Figure 1 shows States by region.

Table 2. Cooperatives marketing selected dairy products, 2002 and 2007[1]

Item	2002	2007	5-year change
	Number of cooperatives		
Bulk raw milk	168	136	-32
Butter	19	19	0
Nonfat dry milk	17	14	-3
Dry whole milk	4	5	+1
Dry buttermilk	11	10	-1
Natural cheese other than cottage cheese	32	31	-1
American cheese	24	18	-6
Italian cheese	6	6	0
Swiss cheese	5	4	-1
Other (specialty) cheeses	11	15	+4
Cottage cheese	9	6	-3
Sour cream	8	8	0
Packaged fluid milk products	16	13	-3
Ice cream	7	6	-1
Ice cream mix	8	6	-2
Yogurt	6	6	0
Condensed milk	14	15	+1
Condensed whey	10	7	-3
Condensed buttermilk	6	4	-2
Dry whey products	13	11	-2
Whey protein concentrates and isolates[2]		6	
Lactose[2]		4	

[1] A cooperative may market several products.

[2] Separately counted for 2007.

Table 3. Cooperative share of milk marketed by producers, 2002 and 2007

Year	Milk from member-producers	Milk from sources other than cooperatives[1]	Total milk handled by cooperatives[2]	United States total	Cooperative share of U.S. Total	
					Member milk	Total milk handled
	Million pounds				Percent	
2002	139,205	5,144	144,349	168,944	82.4	85.4
2007	152,514	3,272	155,786	184,565	82.6	84.4
		Percent-				
5-year change	9.6%	-36.7%	7.9%	9.2%		

[1] Milk from nonmembers and non-cooperative firms.

[2] Handled by physical receipt, bargaining or servicing transactions. Excludes inter-cooperative shipment.

Table 4. Utilization of milk marketed by cooperatives, 2002 and 2007[1]

Year	Milk marketed	Utilization rate
	Million pounds	*---Percent---*
2002		
Sold raw[2]	88,073	61
Processed or manufactured	56,276	39
Total	144,349	100
2007		
Sold raw[2]	98,288	63
Processed or manufactured	57,498	37
Total	155,786	100

[1] Excludes inter-cooperative volume and includes milk from non-member producers and non-cooperative firms.

[2] Includes milk shipped to plants which cooperatives invested in.

MEMBER SUPPLY AND LOCATION

There were 49,675 cooperative member- producers marketing milk in 2007. Most (20,255 producers) were in the East North Central region, followed by the North Atlantic region (12,078 producers) and the West North Central region (10,135 producers). Together, these regions had 85 percent of total member producers, but only 51 percent of member milk (table 5). Following the national trend of decreasing dairy farm numbers, dairy cooperatives reported a 19-percent decline in the number of member-producers in 5 years — from 61,390 to 49,675. The greatest declines were in the two North Central regions, each of which had 4,000 fewer member-producers. Every region of the Nation seemed to be adequately served by dairy cooperatives. The South Atlantic was served by 10 cooperatives, while 62 cooperatives had members in the North Atlantic region. In 2007, the Western region remained the topsource of cooperative milk volume. The 2,736 producers there (5.5 percent of all dairy cooperative members) marketed 58.1 billion pounds of member milk. This re p resented 38 percent of member milk marketed by all cooperatives, up from 35 percent in 2002. The East North Central region accounted for 25 percent of total member milk volume, unchanged from 2002. The North Atlantic and West North Central regions each supplied 13 percent of the total U.S. cooperative member volume.

Table 5. Cooperative member milk by region, number of producers, and milk per producer, 2002 and 2007

Year and region[1]	Cooperatives[2]	Member-producers	Member milk	Milk per producer	Co-op regional share of milk[3]
2002		*Number*	*Million pounds*		*Percent*
North Atlantic	87	12,886	19,826	1.5	69
South Atlantic	8	2,770	8,448	3.0	90
East North Central	53	24,314	34,210	1.4	90
West North Central	53	14,199	17,893	1.3	94
South Central	8	3,617	9,752	2.7	89
Western	24	3,604	49,076	13.6	73
All[4]	194	61,390	139,205	2.3	82.4
2007					
North Atlantic	62	12,078	20,428	1.7	76
South Atlantic	10	2,118	7,350	3.5	94
East North Central	47	20,255	37,675	1.9	91
West North Central	49	10,135	19,192	1.9	97
South Central	11	2,353	9,788	4.2	83
Western	21	2,736	58,081	21.2	76
All[4]	155	49,675	152,514	3.1	82.6

[1] Figure 1 shows States by region.

[2] Cooperatives having member-producers in the region.

[3] Cooperative member milk volume as a percentage of regional volume marketed by producers.

[4] Sum of cooperatives not equal to all-region total because many had member-producers in more than one region.

All regions except the South Atlantic increased milk volumes from cooperative members. The greatest increase was in the Western region, up 9 billion pounds in 5 years. Milk deliveries per member- p roducer were up in all regions in the last 5 years. Nationally, it increased 35 percent from 2.3 million pounds to 3.1 million pounds. Per member delivery was highest in the Western region at 21.2 million pounds, a 56-percent increase from 13.6 million pounds in 2002, and was more than 12 times that of the North Atlantic region in 2007.

PLANT OPERATIONS

Cooperatives owned and operated a total of 193 dairy plants in 2007, 18 fewer than in 2002 (table 6). More than one-third of the plants (66 plants) were in the East North Central region. The region was also the most prominent in almost all categories of dairy plants, except for churning butter, drying milk pro ducts, and making ice cream. The Western region had the most plants for manufacturing butter and dry milk products. Three region had more plants in 2007 than in 2002: the North Atlantic region gained one plant, the South Atlantic region gained five, and the South Central region gained two. On the other hand, the 2 North Central regions and the Western region had fewer plants operated by cooperatives, with the East North Central region showing 17 fewer plants. Seventeen plants served only as milk-receiving stations, compared with 35 in 2002. Manufacturing operations were carried out in 176 plants, unchanged from 2002. Forty-nine cooperative plants packaged fluid milk products, up seven from 2002. The East North Central region had the most, with 14, followed by the West Central region, with 12. Butter-manufacturing plants declined from 25 to 24. Thirty-nine plants manufactured dry milk products (other than dry whey p roducts), down four from 5 years earlier. Many manufacturing operations were devoted to cheesemaking — 34 American, 17 Italian, and 18 other cheeses. The East North Central region had the most cooperative cheese plants in every category. In 2007, 11 cooperative plants made cultured products and the same number made ice cream. Fifty-six plants produced condensed milk and whey pro ducts. Dry whey products were made in 24 cooperative plants. Twenty-six plants performed functions that are not listed in table 6.

DAIRY PRODUCTS MARKETED

This section and the accompanying tables (tables 7 and 8) describe the net volumes of selected dairy products marketed by cooperatives. Comparisons are made between the volumes marketed by cooperatives and total U.S. production.

Butter

Cooperatives churned 1,087 million pounds of butter in 2007, up 14 percent from 2002. Cooperatives' share of U.S. production remained at 71 percent.

Dry Milk Products

A total of 1,444 million pounds of nonfat and skim milk powders (nonfat dry milk, skim milk powders, and dry skim milk for animal) were marketed in 2007 by cooperatives. They had a 96-percent share of U.S. production of these dry milk products. (Skim milk powder is a new category of dry product being reported since 2005. It includes protein standard i z e d and blends.)

Cooperatives marketed 51 percent of the Nation's dry whole milk, up from 47 percent in 2002. Their share of dry buttermilk was 65 percent in 2007.

Cheese

In 2007, cooperatives marketed 2,513 million pounds of natural cheese, excluding cottage cheese, a decline of 15 percent from 2002. As a result, cooperatives' share of U.S. natural cheese p roduction was reduced by 8 percentage points, fro m 34 percent to 26 percent. Among the natural cheese produced by cooperatives, over two-thirds (1,698 million pounds or 68 percent) was American cheese, followed by Italian cheese (743 million pounds or 30 percent). American cheese made by cooperatives was down 20 percent from 2002, while U.S. production increased 5 percent, resulting in a decreased cooperative share of 44 percent, down from 57 percent in 2002. In the same 5 years, cooperative Italian cheese marketing decreased by 3 percent, and cooperative share of U.S. total Italian cheese dipped to 18 percent. Cooperative Swiss cheese production in 2007 was 37 million pounds, 7 percent lower than in 2002. Cooperative share of U.S. production was down to 12 percent. Cooperatives also marketed 34 million pounds of natural cheese other than American, Italian, or Swiss cheese, an 11 - percent increase from

2002. However, this accounted for only 2 percent of U.S. production of other cheeses, unchanged from 2002.

Table 6. Number of dairy plants owned and operated by cooperatives performing various marketing functions, by plant location, 2002 and 2007[1]

	Region[2]						
Marketing function	North Atlantic	South Atlantic	East North Central	West North Central	South Central	Western	Total
2002	*Number*						
Receive and ship milk[3]	11	1	43	41	7	22	125
Churn butter	5	2	6	4	1	7	25
Make dry products	6	2	8	7	3	17	43
Make American cheese	3	0	23	11	0	11	48
Make Italian cheese	2	0	13	4	0	2	21
Make other cheeses	0	0	15	3	0	0	18
Package fluid milk	4	1	14	13	4	6	42
Make cultured products	2	1	5	4	2	1	15
Make ice cream	0	1	9	5	1	4	20
Make condensed products[4]	4	2	26	14	4	16	66
Make dry whey products	2	0	12	10	0	4	28
Other dairy-related activities	6	0	4	6	1	1	18
Total[5]	18	3	83	57	13	37	211[3]
2007							
Receive and ship milk[3]	9	0	42	35	11	26	23
Churn butter	3	2	5	5	1	8	24
Make dry products	6	2	7	6	2	16	39
Make American cheese	4	0	15	9	0	6	34
Make Italian cheese	1	0	11	3	0	2	17
Make other cheeses	1	0	13	4	0	0	18
Package fluid milk	3	5	14	12	9	6	49
Make cultured products	2	0	4	4	0	1	11
Make ice cream	0	0	3	4	1	3	11
Make condensed products[4]	5	3	20	12	2	14	56
Make dry whey products	2	0	10	8	0	4	24
Other dairy-related activities	8	0	3	9	1	5	26
Total[5]	19	8	66	50	15	35	193[3]

[1] All dairy plants, including joint-venture plants operated by co-ops.

[2] Figure 1 shows States by region.

[3] Of plants performing milk receiving and shipping function, 35 were receiving stations only in 2002 and 17 in 2007. Thus, 176 plants in both years were dairy products manufacturing plants.

[4] Includes condensed whey and concentrated milk products.

[5] Sum of plants not equal to totals because some perform more than one function.

Table 7. Selected dairy products marketed by cooperatives, 2002 and 2007

Item	2002	2007	5-year change	
	Thousand pounds			*-Percent-*
Butter	956,211	1,087,012	130,801	14
Dry milk products				
Dry whole milk	22,425	16,322	-6,103	-27
Dry buttermilk	48,104	53,043	4,939	10
Nonfat and skim milk powders[1]	1,352,016	1,444,395	92,379	7
Nonfat dry milk, human		1,357,782		
Skim milk powders		83,790		
Dry skim milk, animal		2,823		
Natural cheese other than cottage cheese	2,948,084	2,512,713	-435,371	-15
American cheese	2,112,011	1,698,485	-413,526	-20
Italian cheese	765,426	742,830	-22,596	-3
Swiss cheese	39,509	36,921	-2,588	-7
Other (specialty) cheeses	31,138	34,477	3,339	11
Dry whey products	1,104,000	1,027,144	-76,856	-7
Dry whey (including modified whey)[2]		739,877		
Whey protein concentrates and isolates		171,077		
Lactose		116,190		
Packaged fluid milk products	3,795,462	4,034,808	239,346	6
Sour cream	132,690	162,827	30,137	23
Cottage cheese	111,424	100,205	-11,219	-10
Condensed whole and skim milk (solids)[3]	629,684	765,221	135,537	22
Condensed buttermilk (solids)	19,000	11,040	-7,960	-42
Bulk fluid cream (butterfat)	326,516	485,536	159,020	49
Bulk fluid skim	294,604	390,331	95,727	32
	Thousand gallons			
Ice cream	38,000	49,268	11,268	30
Ice cream mix	47,000	93,411	46,411	99
Yogurt	4,405	8,129	3,724	85

[1] The three nonfat and skim milk powders are separately counted for 2007. However, nonfat dry milk (human) may be over-counted and skim milk powders and dry skim milk (animal) under-counted, because some cooperatives reported the three products as a lump sum number.

[2] The three dry whey products are separately counted for 2007. Dry whey may include some dry permeate and whey protein concentrates and isolates.

[3] Bulk and canned, including blends and ultrafiltered products.

Table 8. Selected dairy products marketed by cooperatives and shares of U.S. production, 2002 and 2007

hem	2002 co-ops	2002 U.S.	Co-op share	2007 co-ops	2007 U.S.	Co-op share
	Thousand pounds		Percent	Thousand pounds		Percent
Butter	956,21 1	1,355,147	71	1,087,012	1,532,717	71
Dry milk products						
Dry whole milk	22,425	47,411	47	16,322	31,746	51
Dry buttermilk	48,104	54,886	88	53,043	81,386	65
Nonfat and skim milk powders[1]	1,352,0 16	1,603,504	84	1,444,395	1,503,992	96
Natural cheese other than cottage cheese	2,948,0 84	8,547,267	34	2,512,713	9,776,785	26
American cheese	2,112,0 11	3,690,978	57	1,687,485	3,877214	44
Italian cheese	765,42 6	3,470,014	22	742,830	4,198,800	18
Swiss cheese	39,509	254,096	16	36,921	313,689	12
Other (speciality) cheeses	31,138	1,132,179	3	34,477	1,387,082	2
Dry whey products	1,104,0 00	2,116,340	52	1,027,144	2,420,250	42
Dry whey (including modified whey)[2]				739,877	1,231,798	60
Whey protein concentrated and isolates				171,077	432,928	40
Lactose				116,190	755,525	15
Packaged fluid milk products[3]	3,795,4 62	54,216,00 0	7.0	4,034,808	54,619,000	7.4
Sour cream	132,69 0	*n.a.	n.a.	162,827	1,135,468	14
Condensed buttermilk (solids)	19,000	55,875	34	11,040	55,754	20
Ice cream	38,000	1,364,580	3	49,268	1,352,445	4
cem	2002 co-ops	2002 U.S.	Co-op share	2007 co-ops	2007 U.S.	Co-op share
Ice cream milk	47,000	728,701	6	93,411	721,877	13
Yogurt	4,405	70,771	6	8,129	74,722	11

[1] Nonfat dry milk (human) and dry skim milk (animal) for 2002. 2007 data includes skim milk powders.

[2] Dry whey reported by cooperatives may include some dry permeate, and whey protein concentrates and and isolates.

[3] U.S. beverage milk volume adopted from http://www/ers.usda.gov.publications /ldp/LDPTables.htm.

* - not available.

Packaged Fluid Milk Products

Cooperatives marketed 4,035 million pounds of packaged fluid milk products in 2007, up 6 percent from 2002.

However, cooperatives' share of U.S. total beverage milk volume increased only slightly, from 7 percent 5 years earlier to 7.4 percent in 2007.

Ice Cream and Ice Cream Mix

Cooperatives' role in the ice cream business is relatively minor, marketing 49 million gallons in 2007. Their share of the Nation's production increased from 3 to 4 percent. Ice cream made by cooperatives increased 30 percent while U.S. production decreased slightly. Cooperatives reported manufacture of 93 million gallons of ice cream mix or 13 percent of U.S. production in 2007. This market share was more than double the 6 percent share in 2002. In the 5-year period, U.S. production decreased slightly while cooperative volume almost doubled.

Yogurt

The volume of yogurt marketed by cooperatives increased 85 percent since 2002, and market share occupied by cooperatives increased fro m 6 percent to 11 percent.

Dry Whey Products

In 2007, cooperatives marketed 1,027 million pounds of dry whey products, down 7 percent from 2002, while U.S. production increased 14 percent in the same period. The p e rcentage of dry whey products made by cooperatives decreased from 52 percent to 42 percent of U.S. production in 5 years.

Sour Cream and Cottage Cheese

Cooperatives marketed 163 million pounds of sour cream in 2007, an increase of 23 percent over the 5-year period and representing a 14-percent share of U.S. production. Cottage cheese produced by cooperatives decreased by 10 percent to 100 million pounds.

Condensed Milk Products

Condensed whole and skim milk, both bulk and canned, and including blends and ultrafiltered products, totaled 765 million pounds of milk solids. This was a 22-percent increase over 2002. Cooperatives' share of condensed buttermilk dropped to 20 percent of U.S. total in 2007, down from 34 percent.

Cooperative Sizes

The rewere four cooperatives that each handled more than 6 billion pounds of member milk in 2007, the same number as in 2002 (table 9). The four accounted for 49.2 percent of cooperative milk in 2007, the same share as reported for 2002.

The number of cooperatives in the next size g roup (3 billion to 6 billion pounds of milk) increased by one, to eight, by 2007. The milk volume of this g roup occupied a 22.9-percent share of cooperative milk, an increase of 2 percentage points from 2002.

The 2-billion-to-3-billion-pounds group sawadec rease of one cooperative from 2002, and the group's share of cooperative milk decreased by 2.3 percent age points, from 10.5 percent to 8.3 percent.

The most significant change was the 1-billion-to2-billion-pounds group. The number of cooperatives in this group more than doubled from 5 to 11, as did their milk volume. Their share of cooperative milk almost doubled, from 5.1 percent to 10.1 percent in 2007.

The 0.5-billion-to-1-billion-pounds group had eight cooperatives in 2007, five fewer than in 2002. The milk volume accounted for by this group declined the most (by 3.9 billion pounds, or 43 percent), and showed the most substantial decrease in the share of cooperative milk volume, from 6.5 percent to 3.4 percent.

Thirty-two cooperatives were in the group that marketed between 100 million and 500 million pounds of milk in 2007, an increase of two cooperatives. Together, this group had a slightly lower milk volume and had a 4.4-percent share of cooperative milk, down from 4.9 percent in 2002.

The rewere 87 cooperatives in the smallest sized g roup, which marketed less than 100 million pounds of milk in 2007. This was a steep (33 percent) decline fro m 2002, when 129 cooperatives in this group were counted. The group's milk volume declined by 34 percent and held only a 1.8-percent share of cooperative milk, a marked decline from the 2.9-percent share in 2002.

Relative to others, the 1-billion-to-2-billionpounds-of-milk group of cooperatives grew the most in terms of market share. While as a group, the 17 l a rgest sized cooperatives (greater than 2 billion pounds of milk) by and large held steady, decreasing their share of cooperative milk by only 0.3 percentage point from 80.6 percent in 2002 to 80.3 percent in 2007, the smaller sized groups lost more ground.

Concentration Ratio

A measure commonly used to gauge the concentration of a sector (or market) is the share occupied by the four largest firms in the sector.

The share of the largest four dairy cooperatives remained at 49.2 percent of cooperative milk in 2007, as it was in 2002 (tables 9 and 10). Because there was a slight increase in total cooperative milk relative to total U.S. milk marketed (table 3), the largest four dairy cooperatives also had a slightly higher share of the Nation's milk, increasing from 40.5 percent in 2002 to 40.7 percent in 2007.

When the focus was expanded to the largest eight dairy cooperatives, their milk as a share of cooperative milk and as a share of total U.S. milk both went down somewhat from 2002 to 2007.

Meanwhile, the largest 20 dairy cooperatives had a slight decline in their share of cooperative milk. However, their share of the Nation's milk was unchanged.

In terms of milk volume, the relative position of dairy cooperatives to the rest of the industry has been remarkably stable.

Employees

A total of 65 dairy cooperatives (42 percent of all dairy cooperatives) reported having 21,475 full-time and 2,938 part-time employees in 2007 (table 11). These cooperatives marketed 127.4 billion pounds of member milk, or 84 percent of cooperative milk.

(In addition to the main business of marketing milk, some cooperatives also handled farm supplies and/or other commodities. There f o re, not all employees were engaged in the business of handling milk and dairy products.)

Six other cooperatives had only part-time employees. These were small bargaining cooperatives, each employing a part-time employee.

Another 15 cooperatives reportedly had no employees. Again, most of them were small bargaining cooperatives.

Table 9. Size of dairy cooperatives in terms of milk marketed by members, 2002 and 2007

Milk marketed by members	Cooperatives		Share of cooperatives		Member milk		Share of co-op milk	
	2002	2007	2002	2007	2002	2007	2002	2007
	Number		Percent		Million Pounds		Percent	
More than 6 billion pounds	4	4	2.1	2.6	68,499	75,089	49.2	49.2
3 to 6 billon pounds	7	8	3.6	5.2	29,040	34,899	20.9	22.9
2 to 3 billon pounds	6	5	3.1	3.2	14,615	12,504	10.5	8.2
1 to 2 billion pounds	5	11	2.6	7.1	7,120	15,439	5.1	10.1
0.5 to 1 billion pounds	13	8	6.7	5.2	9,101	5,176	6.5	3.4
100 to 500 million pounds	30	32	15.5	20.6	6,761	6,740	4.9	4.4
Less than 100 million pounds	129	87	66.5	56.1	4,063	2,681	2.9	1.8
Total	194	155	100.0	100.0	139,199	152,528	100.0	100.0

In total, these 86 cooperatives re p resented 55 percent of all dairy cooperatives and marketed 86 percent of cooperative milk. The remaining 14 percent of the milk was handled by the 45 percent of dairy cooperatives that did not supply employment information.

Table 10. Share of milk marketed by members of dairy cooperatives, 2002 and 2007

	Category	2002	2007
		Percent	
Share of cooperative volume			
4 largest cooperatives		49.2	49.2
8 largest cooperatives		62.9	62.3
20 largest cooperatives		84.0	83.7
Share of total U.S. volume			
4 largest cooperatives		40.5	40.7
8 largest cooperatives		51.8	51.5
20 largest cooperatives		69.2	69.2

FINANCIAL PERFORMANCE

Complete financial data submitted by 94 dairy cooperatives showed that their total assets for the fiscal year ending in 2007 were $12 billion (table 12). Current assets accounted for $7.3 billion, and fixed assets (net property, plant, and equipment (PPandE) and other assets) were $4.6 billion. Total liabilities for the 94 cooperatives were $8.7 billion. They owed $6.3 billion in current liabilities, and incurred $2.4 billion long-term debt. Members' equity, the balance of assets and liabilities, was $3.3 billion. Most equity (82 percent or $2.7 billion) was allocated to members. On a per- hundred weight-of-milk basis, total assets were $8.41, total liabilities were $6.09, and member equity was $2.32. (Dairy cooperatives typically pay members for their milk twice a month. A large proportion of the current assets and the current liabilities are related to such periodic cash payments to members. This is a unique characteristic of the balance sheet of dairy cooperatives.) The 94 cooperatives marketed 142.9 billion pounds of milk in 2007. They represented 61 percent of all dairy cooperatives and 94 percent of cooperative milk volume. Total income reported by the 94 cooperatives was $44.2 billion (table 13). Of this amount, 87.6 percent ($38.8 billion) was from milk and dairy products sales, 11.6 percent ($5.1 billion) was from supply and other sales, and less than 1 percent was from other income. Substracting total costs and expenses from total income, net margin before taxes was $404 million or 28 cents per hundredweight of milk marketed. This re p resents a before-tax return on equity of 12.2 percent. Some cooperatives reported the value of the milk they bargained for as sales in the income statements, while others did not.

Table 11. Number of employees of 86 dairy cooperatives, 2007

	Cooperatives reporting	Full-time employees	Part-time employees	Member milk
	Number			Million lbs
Having full-time employees	65	21,475	2,938	127,422
Having only part-time employees	6	-	6	296
Having no employee	15	-	-	3,767
Total reported	86	21,475	2,944	131,485
	Percent			
Share of total cooperatives	55%	*n.a.	n.a.	86%

*- not available.

Table 12. Aggregated balance sheet of 94 dairy cooperatives, 2007

Assets:	*--Thousand dollars--*
Current assets	7,258,423
Net PPandE and other assets	4,609,394
Investments in other co-ops	152,067
Assets not categorized	935
Total assets	12,020,819
Liabilities and equity:	
Current liabilities	6,290,839
Long-term and other liabilities	2,409,129
Liabilities not categorized	677
Total liabilities	8,700,645
Members' equity	
Common stock	1,857
Preferred stock	232,595
Allocated equity	2,727,249
Un-allocated equity	358,473
Total equity	3,320,174
Total liabilities and equity	12,020,819
Number of dairy cooperatives reporting	94
Member milk (million pounds)	142,865
Total assets per hundredweight	$8.41
Total liabilities per hundredweight	$6.09
Members' equity per hundredweight	$2.32

For this latter group, an estimated value of the milk that was bargained for was included in the milk and dairy products sales in ord e r for the cooperative sales figures to be consistent. The estimated value is offset by the equal amount of the cost of goods sold and, therefore, does not affect the $404 million net savings reported. Milk and dairy product sales may also include some inter-cooperative transactions; again, this does not affect the reported total net savings of dairy cooperatives as a group.

Table 13. Aggregated income statement of 94 dairy cooperatives, 2007

	Thousand dollars
Milk and dairy product sales1	38,765,715
Supply and other sales	5,128,272
Service receipts, subsidiary and other income	318,880
Patronage refunds received	8,377
Non-recurring gains	15,733
Total income	44,236,977
Cost of goods sold[1]	41,221,393
Expenses	2,606,133
Non-recurring losses	5,161
Total costs and expenses	43,832,687
Net margin before taxes	404,290
Number of dairy cooperatives reporting	94
Member milk (million pounds)	142,865
Net margin before taxes per hundredweight	$0.28
Before-tax return on members' equity	12.2%

[1] Includes the estimated value of milk that was bargained for by some cooperatives but was not reported in their income statements. Both items may also include some inter-cooperative transactions.

In: Dairy Cooperatives Profiles and Research ISBN 978-1-62081-247-1
Editors: A. West and S. I. Kelly

Chapter 3

MEASURING PERFORMANCE OF DAIRY COOPERATIVES*

K. Charles Ling

ABSTRACT

This report revises and simplifies the extra-value approach (developed previously in Research Report 166) for member-producers to evaluate their cooperative's performance. Extra value is defined as net savings after subtracting an interest charge on equity. For comparisons over time and among cooperatives, extra value is expressed as a percentage of operating capital to generate a scale-neutral and mode-neutral, extra-value index for each cooperative or group of cooperatives. The extra-value approach is also used to examine the performance of the surviving cooperatives following mergers and consolidations. The influence of cooperative size on performance is also examined.

PREFACE

In response to inquiries concerning evaluation of the performance of cooperatives and cooperative management, a measurement method called the extra-value approach was proposed in Research Report 166 (published in

* This is an edited, reformatted and augmented version of United States Department of Agriculture, RBS Research Report 212, dated June 2006.

1998). This new evaluation method was developed because the conventional measures of performance-return on equity, return on assets, return on operating capital, net margins on sales, net margins per hundredweight of milk, etc. - do not yield an unequivocal answer to the performance question. Furthermore, cooperatives do not have stock-exchange prices to gauge their performance and market value.

This report revises the approach developed in 1998 to make it clearer and simpler to understand and use extra value as an objective measure to evaluate cooperative performance. The extra-value index is an objective and definitive tool for comparing cooperatives' performance in creating value for member-producers.

For ease of presentation, cooperative codes (#1 through #21) are arbitrarily assigned according to the order a cooperative appears in table 5. Indexes and ratios about cooperative performance are presented after aggregation, and no single cooperative's proprietary data is shown in the report.

HIGHLIGHTS

Without stock market valuation of a cooperative's performance, member-producers often are not sure how well their cooperative performs and whether the cooperative has created or destroyed the value of the cooperative in its operations. They are also perplexed at how the management should be objectively evaluated and rewarded. This is unlike a publicly traded corporation, where the stock price is a timely reflection of the performance and the value of the firm.

Conventional financial ratios-such as return on equity, return on assets, return on operating capital, net margins on sales, net margins per hundredweight of milk, etc.-do not account for the cost of using members' equity in financing a cooperative's operations. The performance of one cooperative may be rated better than that of a second cooperative by some of these measures but worse off by other measures, even if the two cooperatives are exactly the same in every aspect except the way the operations are financed: more by debt and less by equity capital, or vice versa. A definitive measure of cooperative performance called “extra value” was proposed in a previous research report and is further refined and simplified in this study and applied to a set of time-series data to show its usefulness.

The extra-value approach accounts for the total cost of operations, including the cost of using equity, and measures performance in terms of earnings generated above this total operating cost (thus extra value). The cost of using equity is the opportunity cost of equity capital. It is an interest charge on the equity used in the operation at a rate equivalent to the amount the money could earn elsewhere. A positive extra-value figure indicates a cooperative's management has created value for members in its operations; negative extra value means the value of members' investment in the cooperative is diminished.

For the extra-value approach to be an objective performance measure for comparing cooperative operations, extra value can be made neutral to scale and to mode of operations (relative mix of bargaining and/or processing operations) by an index that expresses extra value as a percentage of operating capital. The index shows the rate of creating extra value from using operating capital, which is the financial resource available to the management of a cooperative for operations.

Extra value is defined as net savings after subtracting interest on equity, and operating capital is the sum of fixed (non-current) assets and net working capital.

The analysis used data from dairy cooperatives in USDA's database. The time series were from 1992 through 1996 and from 2000 through 2004. Eleven of the original 28 cooperatives in the 1992-96 period merged or consolidated into four "surviving" cooperatives between 1997 and 1999. For the 1992-96 period, the data of the 11 "predecessor" cooperatives were summed together by assuming that they had already been consolidated into the surviving 4 cooperatives. Therefore, the total number of dairy cooperatives included in this study is 21.

The interest rate used to calculate the cost of using equity was based on the respective year's December average Libor for U.S. dollar loans with a 12-month maturity. Banks in the United States generally will extend loans to a firm with a better-than-average credit rating, at an interest rate of about 200 basis points above the Libor. So 'Libor+2" was the basic rate used to calculate interest on equity.

Equity capital is considered by investors to be riskier than debt. To show the effects of such risk consideration, sensitivity analysis was made using 5 percentage points and 10 percentage points as the risk premium.

The performance of dairy cooperatives is portrayed in three ways to form a composite picture for evaluation: performance categories, changes in performance indexes, and performance rankings.

Cooperatives were placed into five categories according to their performance indexes:

I. Cooperatives that had negative return on equity (one cooperative in the first period; one in the second).
II. Cooperatives that had positive return on equity, but did not generate extra value beyond the cost of using equity capital at basic interest rate (six cooperatives in the first period; one in the second).
III. Cooperatives that generated extra value beyond the cost of using equity capital at basic interest rate but short of reaching 5-percent risk premium (five cooperatives in the first period; 10 in the second).
IV. Cooperatives that generated extra value beyond the cost of using equity capital at basic interest rate plus 5-percent risk premium but short of reaching 10-percent risk premium (three cooperatives in the first period; two in the second).
V. Cooperatives that generated extra values beyond the cost of using equity capital at basic interest rate plus 10-percent risk premium (six cooperatives in the first period; seven in the second).

Most of the cooperatives, 14 (out of 21) in the first period and 19 in the second, generated extra values beyond the cost of using equity capital at basic interest rates (Category III and higher). And six cooperatives in the first period and seven in the second period attained the highest performance, Category V.

The performance indexes of 10 cooperatives showed improvement from the first period (1992-1996) to second period (2000-2004). All performance indexes for each of these cooperatives were more positive, changed from negative to positive, or became less negative in the second period as compared to the first period.

The performance indexes for another three cooperatives were positive in both time periods. However, their respective values declined from the first period to the second period. So they performed less well in the second period as compared to the first.

Likewise, the performance indexes of another six cooperatives all showed declines. Their performance indexes were less positive, changed from positive to negative, or became more negative in the second period. The performance indexes of the remaining two cooperatives showed mixed results.

Cooperatives may be ranked according to four criteria. For constructing the composite picture of cooperative performance, the performance ranking was based on the extra-value index that was calculated using the basic interest rate plus 10-percent risk premium-the most demanding criterion.

A cooperative may be evaluated relative to other cooperatives from the perspective of where it fits in the resulting composite picture.

The composite picture was used to evaluate the impacts on performance by the mergers and consolidations in the late 1990s. Three out of the four surviving cooperatives actually performed better than the sum of their respective predecessor counterparts. They either stayed in the same performance category or climbed to a higher performance category. Their performance indexes showed improvements between the two time periods. And their rankings were all better in the second period.

An interesting question is whether cooperative size affects performance. This may be answered by evaluating the cooperatives as a group-by comparing the weighted-average and simple-average performance indexes.

The weighted-average performance indexes of cooperatives are calculated by adding the financial data across all cooperatives and calculating the performance indexes as if they had been one single organization. Because of the weighting process, larger cooperatives (defined as cooperatives with larger amounts of operating capital, but not necessarily cooperatives with larger volumes of milk or larger numbers of producers) carry more weight and tend to dominate the results.

The simple-average performance indexes treat every cooperative equally, by calculating the performance indexes of each of the 21 cooperatives and then averaging the indexes. The simple averages give an equal weight to each cooperative regardless of size, and no one cooperative has more weight than another to influence the results. Furthermore, 21 is a large enough number of cooperatives, so one or a few cooperatives' performance can not overwhelm the rest.

Comparisons between weighted-average performance indexes and simple averages highlight the performance of larger cooperatives relative to the rest. All weighted-average performance indexes were lower than the corresponding simple averages, suggesting that some of the larger cooperatives did not perform as well as the rest of the cooperatives in either period.

Some of the larger cooperatives also relied less and less on equity and more and more on debt than smaller ones to finance their operations as shown by the differences between the weighted-average equity shares of operating capital and the simple averages. This distorted the comparison based on returns on equity.

In terms of operating efficiency, while the weighted-average extra-value indexes show that the cooperatives as a group did not perform as well in the second period as in the first period, the simple-average extra-value indexes indicate that the performance of the cooperatives on average barely changed between the two periods. This implies that some of the larger cooperatives did not perform as well in the second period when compared to the first period, but the rest of the cooperatives, on average, maintained their level of efficiency in using operating capital.

Member-producers may evaluate their cooperative based on how much extra value the cooperative has created. For comparison over time or with other cooperatives, the extra-value index is an objective and definitive tool and is scale-neutral and mode-neutral.

However, a cooperative is a membership organization as well as a business entity. It has to achieve its business goals, but also has to provide various services to members. The costs and returns of providing such services may not be fully measurable and thus may not be fully reflected in the financial statements and in the extra-value calculation. Furthermore, in a dairy coopeative, the distinction between milk pay prices and premiums on the one hand and profits on the other is not clear-cut. The board and members should consider all these factors when evaluating performance. They are also in the best position to judge the most representative opportunity cost of the cooperative's equity.

The extra-value approach is a worthwhile tool that furnishes some objectivity in evaluating cooperative performance. In the end, what really counts is how satisfied the member-producers are with the cooperative. There is no substitute for a well-informed membership and a vigilant board that understands the complexity of operating a cooperative, both as a business and as a membership organization, to adequately oversee and evaluate its operations.

INTRODUCTION

Without stock market valuation of a cooperative's performance, member-p roducers often wonder how well their cooperative performs and whether it performs at par with or better than other cooperatives. Conventional financial ratios are usually used to provide measures of performance. These include return on equity, return on assets, return on operating capital, net margins on sales, net margins per hundred weight of milk, etc. In the context of cooperatives, however, these performance measures do not account for the cost of using members' equity in financing a cooperative's operations. (This leads some people to characterize equity as free capital-but certainly it is not free, because it has an opportunity cost.) As a result, members cannot be sure whether the cooperative has created or destroyed the value of the cooperative in its operations, and are perplexed at how the management should be objectively evaluated and rewa ded. This is unlike a publicly traded corporation, where the stock price can be a timely reflection of the value of the firm.

The ambiguity of measuring cooperative performance is illustrated in table 1 using two hypothetical dairy cooperatives. Cooperative A and Cooperative B market exactly the same volume of milk and have the same amount of assets, operating capital, sales, cost of goods sold, operating cost, and operating margin. The only diffe rence is the way the operations are financed. Cooperative A's operating capital is financed by $60,000 of debt and $500,000 of member equity, while Cooperative B's operating capital is financed by $310,000 of debt and $250,000 of member equity.

The two cooperatives generate the same operating margin ($60,000), from which interest on debt is subtracted to arrive at net savings. If the interest rate on debts is 5 percent, Cooperative A's net savings is $57,000, or 23 cents per hundredweight of milk, comp a red to Cooperative B's net savings of $44,500, or 18 cents per hundredweight. Cooperative A can claim to have outperformed Cooperative B by all conventional measures of performance (including return on assets, return on operating capital, and net margins on sales). The exception was return on equity-a widely used yardstick of a firm's profitability, which is 11.4 percent for Cooperative Avs. Cooperative B's 17.8 percent.

This example shows that by treating equity as free capital, the conventional performance measure s subject the cooperatives' operating results to open interpretation. A definitive measure of cooperative performance-called "extra value," which includes in its calculation the cost of using equity capital-was proposed in USDA RBS Research Report No. 166. It is further refined and simplified in this study to make it easy to understand and use. (See Box, page 4).

THE EXTRA-VALUE APPROACH

The extra-value approach accounts for the total cost of operations, including the cost of using equity, and measures performance in terms of earnings generated above this total operating cost (thus, the extra value generated).

The cost of using equity is the opportunity cost of equity capital. It is an interest charge on the equity used in the operation at a rate equivalent to the amount the money could earn elsew here. (Economists would call the extra value so created "economic rent," "surplus," or "economic value added." The term "extra value" is more straightforward.)

Table 1. Comparison of two cooperatives using various measures of performance

	Cooperative A	Cooperative B
Milk volume handled (cwt)	250,000	250,000
Total assets	$1,000,000	$1,000,000
Operating capital	560,000	560,000
-financed by Debt	60,000	310,000
-financed by Equity	500,000	250,000
Sales	$5,000,000	$5,000,000
Cost of goods sold	4,500,000	4,500,000
Gross margin	500,000	500,000
Operating cost	440,000	440,000
Operating margin	60,000	60,000
Interest on debt @5 percent	$3,000	$15,500
Net savings	57,000	44,500
Net savings per cwt	0.23	0.18
Return on equity	11.4%	17.8%
Return on assets	5.7%	4.5%
Return on operating capital	10.2%	7.9%
Net margins on sales	1.1%	0.9%
Interest on debt @5 percent	$3,000	$15,500
Net savings	57,000	44,500
Net savings per cwt	0.23	0.18
Interest on equity @5 percent	25,000	12,500
Extra value	32,000	32,000
Extra value index	5.7%	5.7%
Interest on debt @11 percent	$6,600	$34,100
Net savings	53,400	25,900
Net savings per cwt	0.21	0.10
Interest on equity @11 percent	55,000	27,500
Extra value	(1,600)	(1,600)
Extra value index	(0.3%)	(0.3%)

A positive extra value indicates a cooperative's management has created value for members through the cooperative's operations. If a cooperative's operations cannot fully compensate the opportunity cost of using equity capital, thus generating a negative extra value, then the value of members' investment in the cooperative is diminished. Thus, the extra value could be a useful yardstick to measure the performance of the cooperative's management.

Continuing the example in table 1, we will assume the cost of using equity is the same as the 5-percent interest rate on debt. Cooperative A's extra value is calculated to be $32,000, after subtracting a $25,000 interest charge on equity from the $57,000 net savings. Extra value for Cooperative B is also $32,000 after the same calculation by subtracting a $12,500-interest charge on equity from the $44,500 net savings. Each cooperative increases members' value by $32,000.

When the interest rate is 11 percent, neither cooperative can recover all its total cost (including the costs of using equity capital); each cooperative generates a negative extra value of $1,600. Nevertheless, the performance of the two cooperatives is equal-both diminish members' value by $1,600.

This example shows how extra value is a definitive measure of performance. Both Cooperative A and Cooperative B have increased or decreased their members' values by the same amounts and therefore have performed equally well, or poorly, in their respective operations. This is as it should be, because of the assumption that, except for the differences in the proportion of debt and equity financing, every aspect of the cooperatives' operations is the same.

In real life, no two cooperatives are alike. Their sizes and operations may be vastly different. Some dairy cooperatives may have more bargaining operations than processing and marketing. Their margins on sales may be slim, but they do not employ much operating capital in the operation. Other cooperatives may engage more in further processing that uses more operating capital but probably returns higher margins on sales. Still others may have farm supplies and other businesses besides marketing milk (in which case, return per hundredweight of milk would paint a distorted picture of profitability).

For the extra-value approach to be an objective performance measure for comparing cooperative operations, extra value can be made neutral to scale and mode of operations by an index that expresses extra value as a percentage of operating capital. The index shows the rate of creating extra value from using operating capital, which is the financial resource available to the management of a cooperative for operations.

The extra-value index for the two cooperatives in the example is 5.7; that is, 5.7 cents of extra value is generated for every dollar of operating capital employed in the business when the interest rate on equity is 5 percent (table 1). When the interest rate is 11 percent, the extra-value index is minus 0.3; for every dollar of operating capital used in the operations, both cooperatives lost 3-tenths of a cent.

Extra value can be calculated using the information commonly found on any firm's financial statements (except for interest rate on equity, which has to be imputed):

Extra value = net savings minus interest on equity, where interest on equity = members' (non-interest-bearing) equity x interest rate.

The extra-value index is a percentage of operating capital:

Extra-value index = extra value / operating capital x 100, where operating capital = fixed (non-c u r rent) assets + net working capital, while net working capital = current assets minus current liabilities.

In summary, the extra-value approach has the following characteristics:

- Extra value measures how well a co-op uses its operating capital in operations--whether it covers the opportunity cost of using the operating capital. A cooperative is creating value for its members if its extra value is positive. If its extra value is negative, the cooperative is not fully recovering its total input cost, including the opportunity cost of its members' equity capital, and is reducing the cooperative's value.
- An extra-value index can be created by dividing extra value by operating capital. The extra-value index is scale-neutral and can be used to compare performance of cooperatives of diff e rent sizes.
- Operating capital is the financial resource at the disposal of management for the cooperative's operations. Use of operating capital as the denominator in calculating the extra-value index puts various types of cooperatives on an equal footing regardless of their mode of operations, which range from bargaining cooperatives to the most sophisticated processing and marketing cooperatives.

Therefore, extra value and extra-value index meas u re the efficiency with which the management of a cooperative uses operating capital to create value for its member-producers. Operational performance of cooperatives can be compared by using extra-value indexes.

How this report varies from RR 166

For simplicity and clarity in presenting the extra-value approach to cooperative performance analysis, this report uses net savings as the basis (starting point) for calculating extra value. To member-producers, net savings is their cooperative's "bottom line," and is a number that is most easily understood by them and concerns them the most. The definition of operating capital is simply the same as total assets net of current liabilities.

The calculations are thus diff e rent from Research Report 166. Because extra value measures the efficiency of using operating capital in creating value for members, the previous report opted to use net operating margin (before taxes) as the basis for calculating extra value. (Net operating marginis operating margin plus interest and other income and minus interest and other expenses.)

Research Report 166 also excluded investment in other firms from operating capital, with the notion that extra value should capture the cooperative's operating performance but not the performance of other firms in which the cooperative invests.

Therefore, investment in other firms was removed from the cooperative's assets and the corresponding amount was subtracted from members' equity.

Patronage or investment income received was also excluded from the net operating margin because these are not the result of the cooperative's own operations and thus should not play a part in measuring operating performance.

The current report measures the performance of a cooperative's overall operations, while Research Report 166 focused on the performance of the business activities that are under a cooperative's direct operating control. The two measurements, however, are only marginally different.

The upshot is that the extra-value approach may have some variations depending on what emphasis is put on the variables for the calculation of the extra value and the extra-value index. The important thing is that the resulting extra value should be calculated after subtracting the opportunity cost of using equity capital.

SOURCES OF DATA

Financial data - Dairy cooperatives that have complete and continual financial information in the USDA database for its annual top-100 cooperative financial analysis are included in this study (table 2). The time series are from 1992 through 1996 and fro m 2000 through 2004. Using the data from the two 5-year periods makes it possible to show how the performances of the cooperatives pro g ressed over time. (Please note that the database has 1 year of missing data for two cooperatives and 2 years of missing data for another two cooperatives. Considering the amount of information available, the missing data is a minor im perfection.)

Eleven of the original 28 cooperatives in the 1992-96 period merged or consolidated into four "surviving" cooperatives between 1997 and 1999. For the 1992-96 period, the data of the 11 "predecessor" cooperatives were added as if they had already been consolidated into the surviving four cooperatives. Therefore, the total number of dairy cooperatives included in this study is 21. This is necessary for comparing performances over time. The comparisons based on such grouping may not be perfect, but probably are reasonable. (From 1992 to 2004, many smaller cooperatives also merged into the cooperatives included in this study. However, no complete historical financial data for them are available. In any case, their inclusion in this study probably would not have material impacts on the results.)

Interest rates - Ideally, the interest rate used to calculate the cost of using equity should have been the interest rate on a cooperative's debt (the opportunity-cost concept). However, it was difficult to derive a representative rate from a cooperative's various financing activities. An alternative was to use a rate that was based on the respective year's December average British Bankers' Association's London Inter-Bank Offered Rate (BBA Libor for U.S. dollar loans with a 12-month maturity (table 3). Banks in the United States generally will extend loans to a firm with a better-than-average credit rating, at an interest rate of about 200 basis points above the Libor. So 'Libor+2" was the basic rate used to calculate interest on equity (with the implicit assumption that all included cooperatives had better-than-average credit ratings).

Equity capital is considered by investors to be riskier than debt, and they would argue that the imputed interest on equity should be higher than interest on debt to compensate for the risk of investing in the business. Sensitivity analysis could show the effects of such risk consideration by using various rates for the risk premium. This report assumes such risk premiums were 5 and 10 percentage points for the analysis. (The 10 percentage-point premium was chosen because some researchers reported that the historical risk premium of equity was about 9 percent. See Davis et al., page 2.) So "basic interest rate+5" and "basic interest rate+10" were also used in the calculations of extra values and extra-value indexes.

Table 2. Dairy cooperatives included in the study

Names of cooperatives, 1992-96	Names of cooperatives, 2000-04
Agri-Mark, Inc.	Agri-Mark, Inc.
Alto Dairy Cooperative	Alto Dairy Cooperative
Bongards' Creameries	Bongards' Creameries
Cass-Clay Creamery, Inc.	Cass-Clay Creamery, Inc.
Dairylea Cooperative, Inc.	Dairylea Cooperative, Inc.
First District Association	First District Association
Foremost Farms USA	Foremost Farms USA
Md. and Va. Milk Producers Cooperative Assn	Md. and Va. Milk Producers Cooperative Assn
Michigan Milk Producers Association	Michigan Milk Producers Association
Northwest Dairy Assn. (Darigold Farms)	Northwest Dairy Assn. (WestFarm Foods)
O-AT-KA Milk Products Cooperative, Inc.	O-AT-KA Milk Products Cooperative, Inc.
Prairie Farms Dairy, Inc.	Prairie Farms Dairy, Inc.
St. Albans Cooperative Creamery, Inc.	St. Albans Cooperative Creamery, Inc.
Swiss Valley Farms Company	Swiss Valley Farms Company
Tillamook County Creamery Association	Tillamook County Creamery Association
United Dairymen of Arizona	United Dairymen of Arizona
Upstate Milk Cooperatives, Inc.	Upstate Farms Cooperatives, Inc.
Predecessor cooperative(s)	Surviving cooperative
Associated Milk Producers Inc. (All regions)	Associated Milk Producers Inc. (North Central) (1998)1
Land O'Lakes, Inc.	Land O'Lakes, Inc.
Atlantic Dairy Cooperative	(1997)
Dairyman's Co-op Creamery Association	(1998)
Mid-America Dairymen, Inc.	Dairy Farmers of America (1998)
Western Dairymen Cooperative, Inc.	(1998)
Milk Marketing, Inc.	(1998)
Associated Milk Producers Inc. (all regions)	(1998)
California Gold Dairy Products	(1999)
California Milk Producers Association	California Dairies, Inc. (1999)
Danish Creamery	(1999)
San Joaquin Valley Dairymen	(1999)

[1]The number in parenthese indicates the year when a merger or consolidation took place.

In a nutshell: three interest rates were used in the calculation of extra value and extra-value indexes: basic interest rate, which is Libor plus 2 percent; basic interest rate plus 5 percent; and basic interest rate plus 10 percent (table 3). In addition, the conventional returns on equity were calculated for reference purposes.

PERFORMANCE OF DAIRY COOPERATIVES

USDA cannot present each cooperative's proprietary data, so the performance of dairy cooperatives is portrayed in three ways to form a composite picture for evaluation: performance categories, changes in performance indexes, and performance rankings.

Table 3. British Bankers' Association, London Inter-Bank Offered Rate (Libor), actual over 360-day basis, and the interest rates used in this study

Year	Libor, 12-month maturity, December average (%)	Basic rate (Libor+2)	Basic rate+5% risk premium	Basic rate+10% risk premium
1992	4.12984	6.13	11.13	16.13
1993	3.79874	5.80	10.80	15.80
1994	7.57188	9.57	14.57	19.57
1995	5.50452	7.50	12.50	17.50
1996	5.76289	7.76	12.76	17.76
2000	6.23740	8.24	13.24	18.24
2001	2.41674	4.42	9.42	14.42
2002	1.57735	3.58	8.58	13.58
2003	1.49595	3.50	8.50	13.50
2004	3.01515	5.02	10.02	15.02

Performance categories-Cooperatives were placed into five categories according to their performance indexes (table 4):

I. Cooperatives that had negative return on equity (one cooperative in the first period; one in the second).

II. Cooperatives that had positive return on equity, but did not generate extra values beyond the cost of using equity capital at basic interest rate (six cooperatives in the first period; one in the second).

III. Cooperatives that generated extra values beyond the cost of using equity capital at basic interest rate but short of reaching 5-percent risk premium (5 cooperatives in the first period; 10 in the second).

IV. Cooperatives that generated extra values beyond the cost of using equity capital at basic interest rate plus 5-percent risk pre m ium but short of reaching 10-percent risk premium (three cooperatives in the first period; two in the second).

V. Cooperatives that generated extra values beyond the cost of using equity capital at basic interest rate plus 10-percent risk premium (six cooperatives in the first period; seven in the second).

Table 4. Categories of dairy cooperative performance in the two 5-year periods

Performance category	First period average (1992-96)	Second period average (2000-04)
	Co-op code	
I. Cooperatives that had negative return on equity	7	21
II. Cooperatives that had positive return on equity,	4, 8, 9, 10, 11, 12	11
but did not generate extra values beyond the cost of		
using equity capital at basic interest rate		
III. Cooperatives that generated extra values	5, 6, 19, 20, 21	6, 7, 8, 9, 10, 12, ,17,
beyond the cost of using equity capital at basic		18, 19, 20
interest rate		
IV. Cooperatives that generated extra values	2, 3, 18	5, 16
beyond the cost of using equity capital at basic		
interest rate plus 5-percent risk premium		
V. Cooperatives that generated extra values	1, 13, 14, 15, 16, 17	1, 2, 3, 4, 13, 14, 15
beyond the cost of using equity capital at basic		
interest rate plus 10-percent risk premium		

Most of the cooperatives, 14 (out of 21) in the first period and 19 in the second, generated extra values beyond the cost of using equity capital at basic intere strates (Category III and higher). And six cooperatives in the first period and seven in the second period attained the highest performance Category V.

Nine cooperatives (Nos. 2, 3, 4, 5, 7, 8, 9, 10, and 12) moved to a category of better performance in the second period. Cooperative No. 4 improved the most, moving from Category II to Category V.

Meanwhile, four cooperatives (Nos. 16, 17, 18, and 21) moved to a lower performance category.

Four cooperatives (Nos. 1, 13, 14, and 15) maintained the top performing position (Category V) throughout the two 5-year periods. Three other cooperatives (Nos. 6, 19, and 20) remained in Category III and one cooperative (No. 11) in Category II in both periods.

Changes in performance indexes-Changes in the performance indexes of the 21 cooperatives are summarized in table 5. All performance indexes for Cooperative No. 1 were positive in the first period (1992-96) and improved ("+" in the change column) in the second period (2000-04). The cooperative was able to generate net savings that more than covered the opportunity cost of using equity capital that was calculated at the basic interest rate plus 10 percent risk premium. (The extra-value index, or EVI, calculated at basic interest rate +10 percent was positive.)

The net savings of Cooperative No. 2 in the first period was enough to cover the cost of using equity capital at the basic interest rate plus 5 percent risk premium. In the second period, it improved and was able to cover the cost at the basic interest rate plus 10 percent risk premium. All performance indexes showed im p rovement in the second period for Cooperative No. 2.

In the similar manner, Cooperative No. 3 through Cooperative No. 10 improved their performance between the two time periods, as shown by all positive ("+") signs in the change column. All performance indexes for each of these cooperatives were more positive, changed from negative to positive, or became less negative in the second period as compared to the first period.

All three extra-value indexes for Cooperative No. 11 improved (became less negative) over time. However, the return on equity eroded. For Cooperative No. 12, the EVI using the basic interest rate became positive, the EVI using the basic interest rate plus 5-percent became less negative, while the EVI using the basic interest rate plus 10 percent and its return on equity became more negative.

All performance indexes for Cooperative No. 13 through Cooperative No. 15 were positive in both time periods. However, their respective values declined from the first period to the second period ("- " in the change column). So they performed less well in the second period as compared to the first period. Likewise, Cooperative No. 16 through Cooperative No. 21 all showed negative ("-") signs in the change column: Their performance indexes were less positive, changed from positive to negative, or became m o re negative in the second period. By all measures, Cooperative No. 21 was in the doldrums during the recent 5-year period.

Performance rankings-Cooperatives were ranked according to four criteria (return on equity and the three EVIs using respective interest rates) for both the 1992-96 and the 2000-04 periods (table 6). In the first period, rankings of 13 cooperatives varied somewhat depending on the criterion used. For example, Cooperative No. 2 was ranked 8th according to return on equity, while ranked 7th or 9th according to EVIs.

Similar ranking variations applied to 15 cooperatives in the second period.

Cooperative No. 13 was the top performer in both the first and second periods. Cooperative No. 16 was ranked second in the first period, but slipped to 9th place in the second period, while Cooperative No. 1 improved its standing from third to second place.

Some cooperatives, such as Cooperatives No. 4 and No. 7, showed the most improvement in their rankings, while others saw remarkable declines in their rankings.

Table 5. Comparisons of dairy cooperative performance between two 5-year periods

Co-op code	Performance index	1992-96 average	2000-04 average	Change
1	EVI, using basic interest rate	+	+	+
	EVI, using basic interest rate+5%	+	+	+
	EVI, using basic interest rate+10%	+	+	+
	Return on equity	+	+	+
2	EVI, using basic interest rate	+	+	+
	EVI, using basic interest rate+5%	+	+	+
	EVI, using basic interest rate+10%	-	+	+
	Return on equity	+	+	+
3	EVI, using basic interest rate	+	+	+
	EVI, using basic interest rate+5%	+	+	+
	EVI, using basic interest rate+10%	-	+	+
	Return on equity	+	+	+
4	EVI, using basic interest rate	-	+	+
	EVI, using basic interest rate+5%	-	+	+
	EVI, using basic interest rate+10%	-	+	+
	Return on equity	+	+	+
5	EVI, using basic interest rate	+	+	+
	EVI, using basic interest rate+5%	-	+	+
	EVI, using basic interest rate+10%	-	-	+
	Return on equity	+	+	+

Table 5. (Continued)

Co-op code	Performance index	1992-96 average	2000-04 average	Change
6	EVI, using basic interest rate	+	+	+
	EVI, using basic interest rate+5%	-	-	+
	EVI, using basic interest rate+10%	-	-	+
	Return on equity	+	+	+
7	EVI, using basic interest rate	-	+	+
	EVI, using basic interest rate+5%	-	-	+
	EVI, using basic interest rate+10%	-	-	+
	Return on equity	-	+	+
8	EVI, using basic interest rate	-	+	+
	EVI, using basic interest rate+5%	-	-	+
	EVI, using basic interest rate+10%	-	-	+
	Return on equity	+	+	+
9	EVI, using basic interest rate	-	+	+
	EVI, using basic interest rate+5%	-	-	+
	EVI, using basic interest rate+10%	-	-	+
	Return on equity	+	+	+
10	EVI, using basic interest rate	-	+	+
	EVI, using basic interest rate+5%	-	-	+
	EVI, using basic interest rate+10%	-	-	+
	Return on equity	+	+	+
11	EVI, using basic interest rate	-	-	+
	EVI, using basic interest rate+5%	-	-	+
	EVI, using basic interest rate+10%	-	-	+
	Return on equity	+	+	-
12	EVI, using basic interest rate	-	+	+
	EVI, using basic interest rate+5%	-	-	+
	EVI, using basic interest rate+10%	-	-	-
	Return on equity	+	+	-
13	EVI, using basic interest rate	+	+	-
	EVI, using basic interest rate+5%	+	+	-
	EVI, using basic interest rate+10%	+	+	-
	Return on equity	+	+	-
14	EVI, using basic interest rate	+	+	-
	EVI, using basic interest rate+5%	+	+	-
	EVI, using basic interest rate+10%	+	+	-
	Return on equity	+	+	-
15	EVI, using basic interest rate	+	+	-
	EVI, using basic interest rate+5%	+	+	-
	EVI, using basic interest rate+10%	+	+	-
	Return on equity	+	+	-
16	EVI, using basic interest rate	+	+	-
	EVI, using basic interest rate+5%	+	+	-

Co-op code	Performance index	1992-96 average	2000-04 average	Change
	EVI, using basic interest rate+10%	+	-	-
	Return on equity	+	+	-
17	EVI, using basic interest rate	+	+	-
	EVI, using basic interest rate+5%	+	-	-
	EVI, using basic interest rate+10%	+	-	-
	Return on equity	+	+	-
18	EVI, using basic interest rate	+	+	-
	EVI, using basic interest rate+5%	+	-	-
	EVI, using basic interest rate+10%	-	-	-
	Return on equity	+	+	-
19	EVI, using basic interest rate	+	+	-
	EVI, using basic interest rate+5%	-	-	-
	EVI, using basic interest rate+10%	-	-	-
	Return on equity	+	+	-
20	EVI, using basic interest rate	+	+	-
	EVI, using basic interest rate+5%	-	-	-
	EVI, using basic interest rate+10%	-	-	-
	Return on equity	+	+	-
21	EVI, using basic interest rate	+	-	-
	EVI, using basic interest rate+5%	-	-	-
	EVI, using basic interest rate+10%	-	-	-
	Return on equity	+	-	-

Evaluating the performance of dairy cooperatives - The composite picture of cooperative performance emerging from examining cooperatives' performance categories, changes in performance indexes, and performance ranking are presented in table 7.

The 2000-04 period saw 7 cooperatives in performance Category V, 2 in Category IV, 10 in Category III, and 1 each in Categories II and I. This resulted from seven cooperatives moving up one level from the 1992-96 period, one moving up two levels, and another one moving up three levels. On the other hand, two cooperatives moved down one level and another two moved down two levels. Eight cooperatives stayed in their same respective categories throughout the two time periods.

Eleven cooperatives showed improvement in their extra-value indexes, nine showed deterioration, and one had mixed results. Eleven cooperatives also moved higher in the performance ranking, nine moved lower and the top-ranked cooperative stayed put. (In these two instances, the number of cooperatives appears to be the same, but they are not necessarily the same cooperatives.) The performance ranking was based on the EVI that was calculated using basic interest rate plus 10 percent risk premium - the most demanding criterion.

Table 6. Comparisons of dairy cooperative rankings between two 5-year periods

	1992-96 average				2000-04 average			
Co-op code	Return on equity	EVI, using basic interest rate (Libor+2)	EVI, using basic interest rate+ 5%	EVI, using basic interest rate+ 10%	Return on equity	EVI, using basic interest rate (Libor+2)	EVI, using basic interest rate+ 5%	EVI, using basic interest rate+ 10%
1	3	3	3	3	2	2	2	2
2	8	7	7	(9)	4	3	3	4
3	9	8	8	(8)	3	5	4	3
4	20	(20)	(20)	(18)	7	8	8	7
5	10	10	(10)	(10)	8	7	7	(8)
6	15	14	(15)	(14)	15	13	(12)	(13)
7	(21)	(21)	(21)	(21)	10	10	(10)	(14)
8	18	(18)	(16)	(16)	14	15	(14)	(11)
9	16				11	11	(15)	(17)
10	19	(19)	(18)	(17)	13	12	(11)	(10)
11	17		(17)	(19)	20	(20)	(20)	(20)
12	13	(15)	(13)	(13)	19	19	(19)	(19)
13	1	1	1	1	1	1	1	1
14	6	5	5	6	6	4	5	5
15	4	4	4	4	5	6	6	6
16	2	2	2	2	9	9	9	(9)
17	5	6	6	5	12	14	(13)	(12)
18	7	9	9	(7)	17	18	(18)	(16)
19	12	12	(12)	(12)	16	16	(16)	(15)
20	11	11	(11)	(11)	18	17	(17)	(18)
21	14	13	(14)	(15)	(21)	(21)	(21)	(21)

*Parentheses indicate the cooperative had negative return (loss) for the particular entry.

A cooperative may be evaluated relative to other cooperatives from the perspective of where it fits in the picture. For example, among the seven top performing cooperatives (performance Category V) in the 2000-04 period, members of Cooperative No. 13 probably would be satisfied that their cooperative retained the top rank. But they may also notice that ranking alone does not tell the whole story. They may wonder what had transpired during the past 5 years that their cooperative's performance indexes weakened, while Cooperatives Nos. 1, 3, and 2 were catching up.

Members of Cooperative No. 14 may also be left wondering, because the cooperative's performance indexes also declined, although its ranking moved up. Members of Cooperative No. 15 will certainly be puzzled, because their cooperative managed to remain in the top performing group, even though both performance indexes and ranking sagged.

Table 7. A composite portrait for evaluating performance of dairy cooperatives, the 2000-04 period compared with the 1992-96 period

Co-op code	Performance category 2000-04	Change in performance category	Changes in performance indices (EVIs)[1]	2000-24 ranking based on EVI using basic interest rate+10%[2]
13	V	same	-	1st (+/-)
1	V	same	+	2nd (+)
3	V	up 1 level	+	3rd (+)
2	V	up 1 level	+	4th (+)
14	V	same	-	5th (+)
15	V	same	-	6th (-)
4	V	up 3 levels	+	7th (+)
5	IV	up 1 level	+	8th (+)
16	IV	down 1 level	-	9th (-)
10	III	up 1 level	+	10th (+)
8	III	up 1 level	+	11th (+)
17	III	down 2 levels	-	12th (-)
6	III	same	+	13th (+)
7	II	up 2 levels	+	14th (+)
19	III	same	-	15th (-)
18	III	down 1 level	-	16th (-)
9	III	up 1 level	+	17th (+)
20	III	same	-	18th (-)
12	III	up 1 level	mixed	19th (-)
11	II	same	+	20th (-)
21	I	down 2 levels	-	21st (-)

[1]"+" means a cooperative's EVIs were more positive, changed from negative to positive, or became less negative in the 2000-04 period as compared to 1992-96 period. "-" means the opposite.

[2]Next to the rank, (+) means the cooperative's ranking improved from the 1992-96 period to the 2000-04 period, while (-) means the ranking dropped. (+/-) means no change.

On the other hand, members of Cooperative No. 4 probably would be elated that their cooperative's performance improved remarkably, climbing three levels from Category II to reach Category V, where it ranked 7th.

In conducting the performance evaluation, each cooperative will have to ask a host of questions: What happened? What worked? What did not? What im p rovements need to be made? How to achieve the goals? And so on.

The answers to these questions are certainly going to vary, depending on a cooperative's particular situation. If a cooperative did not turn in a stellar performance during the past 5 years, however, it probably should not be blamed on "market conditions," which, after all, were faced by all dairy cooperatives. But not every cooperative experienced the same ill effects. There was an almost even split in the number of cooperatives that performed better or worse (11 to 9, judged either by performance indexes or by performance ranking) than in the 1992-96 period.

Merger and Consolidation

In the interim years between the first and second periods, 11 cooperatives merged or consolidated to form 4 surviving cooperatives (table 2). An interesting question: did the merger and consolidation improve the performance of the cooperatives?

The performance of the four surviving cooperatives was compared to that of the corresponding groupings of the predecessor cooperatives. Three out of the four surviving cooperatives actually did perform better than the sum of their respective predecessor counterparts. They either stayed in the same performance category or climbed to a higher performance category. Their performance indexes showed improvements between the two time periods. And their rankings were all better in the second period.

Did Size Matter?

One way to answer the question of whether cooperative size affected performance is by evaluating the cooperatives as a group-by comparing the weighted-average and simple-average performance indexes.

The weighted-average performance indexes of cooperatives are calculated by adding the financial data across all cooperatives and calculating the performance indexes as if they had been one single organization. Because of the weighting process, larger cooperatives (defined as cooperatives with larger amounts of operating capital, but not necessarily cooperatives with larger volumes of milk or larger numbers of producers) carry more weights and tend to dominate the results.

The simple-average performance indexes treat every cooperative equally, by calculating the performance indexes of each of the 21 cooperatives and then averaging the indexes. The simple averages give an equal weight to each cooperative regardless of size. No one cooperative has more weight than another to influence the results. Furthermore, 21 is a large enough pool of cooperatives that one, or even a few, cooperatives' performance cannot overwhelm the rest.

If the weighted-average performance indexes show that the cooperatives as a group performed better than indicated by the simple averages, it may be inferred that larger cooperatives used operating capital better than smaller ones, and vice-versa.

Weighted averages-Combining the 21 dairy cooperatives as if one single entity, the return on equity would be 10 percent in 1992 (annual weighted average,). The returns on equity for the other years were all in double digits, except in 2004, when the return was 9.6 percent.

The average return on equity during the first period (1992-96) was 13.1 percent, while it was 10.9 percent for the second period (2000-04).

The net savings of the group was able to pay for the opportunity cost of using equity capital at the basic interest rate. This savings generated an extra value that was 2.4 percent of the combined net operating capital in 1992 (EVI was 2.4 percent, table 8). All EVIs using the basic interest rates as the opportunity costs were positive during the 10 years of this study, with the first period having an average EVI of 4.9 percent and the second period, 2.7 percent.

If the opportunity cost of using equity capital had a risk premium of 5 percent above the basic rate, then the group was not able to cover the cost of using the equity capital in 1992 (when EVI was -0.7 percent) or in 2004 (when EVI was -0.5). EVI was positive for the other 8 years. The average EVI for the first period was 1.4 percent and 0 for the second period.

The combined net savings of the cooperatives was not able to cover the opportunity cost of using equity capital plus the 10 percent risk premium during any of the 10 years. EVIs were all negative. The 5-year average EVI for the first period was -2 percent, while the second period average was -2.7 percent.

Judged as if one single group, the 21 cooperatives did not perform as well in the second period as in the first period. The return on equity decreased 2.2 percentage points from 1992-96 to 2000-04, and the thre e EVIs also declined, respectiv ely, by 2.1, 1.4 and 0.7 percentage points (table 8). This occurred as the equity shareof operating capital was reduced by 14 percentage points and the cooperatives as a group relied proportionately more on debts to finance their operations.

Simple averages-By averaging the individual performance of the 21 dairy cooperatives, the 5-year average return on equity was 16.8 percent during the first period (1992-96) and decreased to 14.3 percent during the second period (2000-04).

The average EVI at the basic interest rate was 7.3 percent in 1992 (table 9). All average EVIs at the basic interest rates were positive during the 10 years, averaging 8.3 percent for the first period and 8.1 percent for the second.

If the opportunity cost of using equity capital had a risk premium of 5 percent above the basic rate, average EVIs were still positive for all 10 years. The average EVIs, respectively, for the two 5-year periods, were 4.5 percent and 4.4 percent.

However, if the risk premium was 10 percent above the basic interest rate, the results were mixed.

The EVIs were positive for some years, but negative for others. The 5-year average EVI was 0.8 percent for both periods.

The three average EVIs that were calculated at the respective interest rates changed very little fro m the first 5-year period to the most recent (table 9), suggesting that the performance of the cooperatives barely changed. So, while the average return on equity declined in recent years, the extra-value indexes show that the efficiency in using operating capital remained largely unchanged.

The average equity share of operating capital also changed little between the two periods, decreasing by only 3 percentage points (table 9). The dairy cooperatives on average seemed to maintain their equity close to a level that is proportional to the requirement of operating capital.

So, did size matter? Comparisons between weighted-average performance indexes and simple averages highlight the performance of larger cooperatives relative to the rest. All weighted-average performance indexes were lower than the corresponding simple averages, suggesting that some of the larger cooperatives did not perform as well as the rest of the cooperatives in either period (tables 8 and 9). Some of the larger cooperatives also relied less on equity and m o re on debt than smaller ones to finance their operations-as shown by the diff e rences between the weighted-average equity shares of operating capital and the simple averages.

The returns on equity, either weighted average or simple average, dipped by similar proportions fro m the first period to the second. This seems to indicate that the profitability of larger and smaller cooperatives declined by about the same proportion. But notice that the weighted-average equity share of operating capital declined 14 percentage points between the two periods, while the simple average decreased only 3 percentage points. Apparently, some of the larger cooperatives relied far more heavily on debt and less on equity to finance their operations in the second period and thus enhanced the return on equity. Otherwise, the weighted-average return on equity would have shown steeper decline than the simple average in the second period. (This is another example of why return on equity is not a reliable performance measure.)

In terms of operating efficiency, while the weighted-average EVIs show that the cooperatives as a group did not perform as well in the second period as in the first period, the simple-average EVIs indicate that the average performance barely changed between the two periods. This implies that some of the larger cooperatives did not perform as well in the second period as in the first period. But the rest of the cooperatives, on average, maintained their level of efficiency in using operating capital.

Table 8. Performance of 21 dairy cooperatives as a group, annual weighted averages

Year	Return on equity	EVI (i=basic)	EVI (i=basic+5%)	EVI (i=basic+10%)	Equity share of operating capital
			Percent		
1992	10.0	2.4	(0.7)	(3.8)	62
1993	12.7	4.8	1.2	(2.3)	71
1994	15.1	4.8	1.2	(2.3)	71
1995	12.8	6.2	2.8	(0.6)	69
1996	14.8	6.1	2.6	(0.8)	69
Average[1]	13.1	4.9	1.4	(2.0)	68
2000	11.8	3.5	0.5	(2.5)	60
2001	11.2	2.7	0.1	(2.5)	52
2002	11.1	2.8	0.0	(2.7)	55
2003	10.8	2.6	0.0	(2.5)	51
2004	9.6	2.1	(0.5)	(3.1)	52
Average[1]	10.9	2.7	0.0	(2.7)	54
Change	(2.2)	(2.1)	(1.4)	(0.7)	(14)

[1]Five-year simple average.

Table 9. Performance of 21 dairy cooperatives as a group, annual simple averages

Year	Return on equity	EVI (i=basic)	EVI (i=basic+5%)	EVI (i=basic+10%)	Equity share of operating capital
			Percent		
1992	13.7	7.3	3.5	(0.3)	75
1993	17.4	10.4	6.6	2.7	77
1994	16.7	6.5	2.7	(1.2)	76
1995	16.9	8.2	4.5	0.8	76
1996	19.2	9.3	5.5	1.7	76
Average[1]	16.8	8.3	4.5	0.8	76
2000	15.5	6.8	3.0	(0.7)	75
2001	15.8	9.5	5.9	2.2	73
2002	14.2	9.4	5.8	2.2	72
2003	11.0	6.3	2.7	(0.9)	72
2004	14.8	8.4	4.8	1.2	72
Average[1]	14.3	8.1	4.4	0.8	73
Change	(2.5)	(0.3)	(0.1)	0.0	(3)

[1]Five-year simple average.

The comparisons in this section offer some interesting general observations. But without presenting individual cooperatives' data, it is difficult to make a definitive conclusion about cooperative size and performance. Suffice it to say that some of the larger cooperatives did not perform as well as other cooperatives, that the performance of some of the larger cooperatives (not necessarily the same ones) slipped over time, and that some of the larger cooperatives relied proportionately less and less on member equity for financing operations.

Conclusion: No Substitute for Board and Member Vigilance

The extra-value approach is a useful tool for member-producers to evaluate the performance of their cooperative:

- Extra value measures whether and by how much the cooperative's net savings exceeds the opportunity cost of member equity.

- Extra-value index measures the rate at which the extra value is generated given the operating capital used in the cooperative's operations.
- Being scale-neutral and mode-neutral, the extra-value index is an objective and definitive tool for comparing performance over time and among cooperatives.

However, a cooperative is a membership organization as well as a business entity. It has to achieve its business goals, but also has to satisfy its members' objectives. Besides expecting good returns by marketing milk through the cooperative as an assured market, dairy farmers also look to the cooperative to provide field and other services (e.g., assist with p roduction problems, assist with inspection problems, sell milking supplies and equipment, provide information on price and availability of hay and heifer replacements, provide marketing and outlook information, provide insurance programs (life, health, disaster), provide retirement programs, negotiate hauling rates, collect and ensure payment fro m buyers, check weights and tests, represent members' interests in government, regulatory and public a ffairs, and so on). The returns of providing such member services may not be fully measurable and thus may not be fully reflected in the financial statements. The extra-value index, like any other financial ratios, does not capture the value of member benefits that are not quantified. The board and members should be cognizant of the value of such benefits in addition to financial returns, when evaluating their cooperative's performance.

In a dairy cooperative, the distinction between milk pay prices and premiums on the one hand and profits on the other is not clear-cut. If a dairy cooperative pays members high prices and premiums for milk, it may report low margins, or even incur losses. On the other hand, a cooperative may pay lower milk prices and report hefty margins. The two cooperatives may perform equally well, although their financial results show otherwise. However, if milk procurement is competitive, this consideration may have only minimal, or no, effect. The board and members should take their cooperative's pricing policies into consideration in evaluating its performance.

The opportunity cost (including risk premium) of equity capital is specific to each individual cooperative, which most likely is diff e rent from the interest rates used in this report. Also, each member's financial situation may be different and their opportunity cost of capital may vary from one to another. The board and the members are in the best position to judge the most re p resentative interest rates to use in the extra-value calculation.

The extra-value approach is a worthwhile tool that furnishes some objectivity in evaluating cooperative performance. In the end, what really counts is how satisfied the member- p roducers are with the cooperative. There is no substitute for a well-informed membership and a vigilant board that understands the complexity of operating a cooperative, both as a business and as a membership organization, to adequately oversee and evaluate its operations.

This report applies the extra-value approach to dairy cooperatives as an example. The approach should be equally applicable to cooperatives marketing other commodities. For comparing the performance of cooperatives across commodities, however, care should be taken to consider the characteristics of the various commodity sectors. (For example, the amount of operating capital required to market a unit of one commodity may be very diff e rent from that required for another commodity, and the margins on sales also may be different.) Although the extra-value index is scale-neutral and mode-neutral, it may, or may not, be commodity-neutral because of the characteristics of the respective sectors.

REFERENCES

Davis, Evan, C. Gouzouli, M. Spence, and J. Star. "Measuring the Performance of Banks," *Business Strategy Review*, Autumn 1993, Vol. 4, No. 3.

Ling, K. Charles and Carolyn Liebrand. *A New Approach to Measuring Dairy Cooperative*

Performance, Research Report 166, Rural Business-Cooperative Service, U.S. Department of Agriculture, September 1998.

The British Bankers' Association. *Historic Libor Rates*, http://www.bba.org.uk/public/libor/.

INDEX

A

access, vii, 1, 2, 5, 6, 7, 8, 9, 12, 40, 43
accounting, 16, 51
age, 64
agencies, 10, 12, 13, 21
aggregation, 72
agricultural market, 19, 44
agriculture, vii, 3, 10, 11
appraisals, 52
assessment, 7, 44
assets, viii, 6, 7, 24, 25, 30, 38, 42, 43, 45, 49, 52, 67, 68, 72, 73, 76, 77, 78, 80, 81
attitudes, 30
audits, 22

B

balance sheet, 24, 25, 45, 67, 68
bankruptcy, 32
bargaining, 4, 11, 23, 27, 31, 32, 41, 56, 66, 73, 79, 80
base, 2, 5, 27, 32, 46
basis points, 73, 82
benefits, vii, 1, 5, 38, 46, 97
beverages, 34
blends, 59, 61, 64
board members, 22
bones, 13
brand loyalty, 34
breeding, 15
business function, 17, 27
business model, vii, 1, 4, 16, 27, 40, 41, 42, 46
business partners, 32
businesses, 3, 5, 10, 26, 33, 79
buyers, 20, 21, 22, 23, 27, 31, 97

C

capital employed, 80
capital intensive, 42
case study, 2, 3, 5, 9, 44
cash, 24, 26, 27, 31, 32, 39, 67
challenges, vii, 2, 3, 7, 13, 44
cheese, viii, 19, 23, 24, 28, 31, 32, 36, 39, 49, 51, 53, 55, 58, 59, 60, 61, 62, 64
chemicals, 45
clarity, 81
commerce, 38
commercial, 11, 13, 36
commodity, 19, 31, 32, 35, 44, 98
communication, 22
community, 46
comparative advantage, 23, 28
compensation, 20
competition, vii, 1, 10, 12, 13, 18, 28, 46
complexity, 76, 98
composition, 17, 23, 35

conflict, 38
consolidation, 83, 92
consulting, 20
consumers, 34
consumption, 37
cooperation, 3, 9, 10, 11, 13, 27, 50
coordination, vii, 1, 10, 17, 18, 29
cost, vii, 1, 4, 12, 14, 16, 18, 29, 32, 34, 44, 46, 68, 72, 73, 74, 76, 77, 78, 79, 80, 81, 82, 84, 85, 86, 93, 94, 96, 97
covering, 11
credit rating, 73, 82
crops, 45
customers, 24, 26

D

dairy industry, vii, 3, 19, 23, 33, 36
database, 73, 82
debts, 38, 77, 94
deficit, 4, 15, 18
Department of Agriculture, 1, 48, 71
destiny, viii, 2, 46
directors, 14, 17, 18, 21, 22, 28, 30, 39, 43
disaster, 20, 97
dissatisfaction, 30
distribution, 4, 7, 13, 22, 37, 43
distribution of income, 4
diversity, 16
division of labor, 23
dominance, 23
drought, 35
drying, 51, 58

E

earnings, 6, 16, 25, 26, 27, 30, 38, 40, 43, 73, 78
economic development, viii, 2, 43, 46
economic rent, 78
Economic Research Service, 45, 47, 48
economic theory, vii, 1, 2, 3, 4, 9, 27, 39
economies of scale, 8, 41, 45
egg, 6, 40
election, 17
employees, viii, 20, 49, 52, 66, 68
employment, 66
English Language, 46
entrepreneurs, viii, 2, 46
entrepreneurship, 10
environment, 52
equipment, 7, 20, 31, 32, 34, 67, 97
equities, 4, 5, 24, 25, 26, 27, 29, 30, 31, 43
ERS, 35, 36
ethanol, 42
evolution, 9, 52, 54
exercise, vii, 1, 11, 21
expenditures, 31
expertise, 33, 34
export market, 33, 36, 37
exporter, 35, 37
exporters, 37
exports, 33, 35, 36, 37
exposure, 39

F

families, 20
farmers, vii, 1, 3, 4, 10, 11, 12, 13, 14, 15, 17, 18, 19, 20, 21, 23, 27, 28, 30, 31, 32, 34, 44, 50, 97
farms, vii, 1, 3, 4, 14, 15, 16, 17, 18, 19, 21, 28, 29, 30, 31, 36, 37, 52
fat, 35, 36
feedstock, 42
financial, viii, 2, 4, 7, 18, 24, 26, 29, 31, 32, 34, 37, 43, 44, 49, 50, 52, 67, 72, 73, 75, 76, 79, 80, 82, 92, 97
financial condition, 34
financial data, viii, 7, 24, 44, 49, 52, 67, 75, 82, 92
financial reports, 26, 43
financial resources, 4, 24, 29, 34, 37
financial support, 32
flexibility, 38
fluid, viii, 19, 22, 23, 24, 28, 36, 38, 41, 49, 51, 55, 58, 60, 61, 62, 63
food, 32, 45
force, 2

formation, 40
fruits, 45
funding, 46
funds, 29, 30, 38

G

genetics, 15
geography, 11
globalized marketplace, 8, 40
goose, 6, 40
governance, vii, 3, 4, 7, 8, 9, 17, 18, 22, 27, 40, 42, 43
grouping, 82
growth, 8, 33, 37, 38
guidance, 12, 17, 28

H

health, 20, 97
history, 47
horizontal integration, vii, 1, 17, 18, 29
host, 92
human, 61, 62

I

improvements, 75, 92
income, 15, 16, 26, 30, 39, 67, 69, 81
independence, 15, 18, 28
Independence, 47
individuality, 15, 18, 28
industry, 52, 54, 65
ingredients, 32, 36
institutions, 3, 12, 13, 18, 28, 31
interest rates, 74, 84, 87, 93, 94, 97
international trade, 36
investment, 6, 7, 8, 15, 25, 26, 32, 33, 34, 36, 37, 38, 39, 41, 42, 43, 54, 73, 79, 81
investments, 44
investors, 7, 20, 35, 38, 43, 73, 82
issues, vii, 2, 5, 6, 7, 8, 9, 20, 32, 33, 36, 38, 39, 40, 41, 42, 46

J

joint ventures, 5, 38

K

Kenneth Galbraith, 11

L

laws, 2, 7, 9, 28, 43
lending, 31
limited liability, 5, 37
livestock, 15
loans, 73, 82
logistics, 35
long-term debt, 5, 8, 24, 40, 67
loyalty, 31

M

machinery, 45
magnitude, 30
major decisions, 22
majority, 38
management, 17, 20, 22, 33, 34, 40, 71, 72, 73, 77, 79, 80
manufacturing, viii, 12, 13, 23, 32, 40, 41, 45, 49, 51, 55, 58, 60
market access, 10
market discipline, 3, 13
market economy, vii, 2
market position, 11, 34
market share, 19, 23, 24, 51, 54, 63, 65
marketplace, 9, 10, 11, 12, 17, 18, 27
matter, iv, 5, 94
measurement, 71
measurements, 81
membership, 8, 16, 21, 22, 25, 30, 31, 32, 33, 39, 40, 46, 76, 97, 98
merchandise, 45
mergers, viii, 71, 75
mission, vii, 3, 23, 27

Missouri, 47
modernization, 32, 41
multinational firms, 36

N

narratives, 9, 18
negotiating, 4, 20
neutral, viii, 71, 73, 76, 79, 80, 97, 98
niche market, 23
niche marketing, 23

O

objectivity, 76, 98
Oceania, 35
oil, 45
opportunities, 6, 35
opportunity costs, 93
organize, vii, 1, 5, 10, 17, 19, 40
ownership, 21, 29, 35, 38

P

participants, vii, 1, 2, 3, 13, 17, 23
petroleum, 42, 45
plants, viii, 7, 19, 22, 23, 26, 31, 32, 34, 37, 41, 46, 49, 51, 52, 53, 54, 55, 56, 58, 60
policy, 22
population, 16
poultry, 45
precedents, 46
preparation, iv
present value, 2, 30
President, 46
pricing policies, 97
principles, vii, 1, 13, 14, 18, 29
procurement, 23, 97
producers, viii, 10, 11, 12, 17, 19, 28, 38, 39, 42, 44, 49, 50, 51, 53, 54, 56, 57, 71, 72, 75, 76, 81, 93, 96
profit, 15, 16, 17, 35, 38, 43
profitability, 77, 79, 95
property rights, 6, 42
proportionality, vii, 1, 14, 21
proposition, 44
prosperity, 36
public policy, 46
purchasing power, 12

Q

quality control, 23
quality of service, 12

R

real estate, 7, 43
reality, 18, 24, 33, 40
reasoning, 15
recession, 35
recommendations, iv
regeneration, 11
relevance, 9
rent, 80, 81, 97, 98
requirements, 23, 31
researchers, 83
response, 11, 39, 71
restructuring, 54
retained earnings, 2, 4, 6, 24, 26, 27, 42, 43
retirement, 30, 32, 41, 97
revenue, vii, 3
risk, 10, 24, 32, 39, 42, 73, 74, 75, 82, 84, 85, 86, 90, 93, 94, 97
risk management, 32
risks, 6, 24, 28, 42, 46
rules, 7, 17, 43
rural areas, 44

S

safety, 32
savings, viii, 26, 29, 43, 68, 71, 73, 77, 78, 79, 80, 81, 86, 93, 96
scale economies, 3, 13, 46
scope, 2, 10, 17, 21, 27, 52
sensitivity, 73

shareholders, 15, 16, 17, 38, 42
showing, 32, 58
signals, 11
signs, 86
solution, vii, 2, 5, 46
state, 7, 38, 44
statistics, 50
stimulus, 12
stock, 5, 6, 25, 27, 37, 38, 42, 72, 76
stock price, 72, 77
stockholders, 15
stress, 32, 34
structure, 3, 4, 7, 9, 13, 16, 18, 27, 43
structuring, 5
supplier, 36, 37
suppliers, 37
surplus, 4, 14, 15, 18, 23, 26, 78

T

target, 27, 36, 37
taxes, viii, 30, 50, 52, 67, 69, 81
techniques, 11
technology, 7, 24, 28, 32, 37, 40, 42
tension, 8, 29
territory, 33
time periods, 50, 74, 75, 86, 89, 92
time series, 73, 82
total costs, 67
trade, 4, 20, 22, 23, 27, 33, 36
traits, 16
transactions, 4, 14, 15, 26, 43, 56, 69
transportation, 21
treatment, 38

U

U.S. Department of Agriculture, 8, 46, 48, 98
United, 1, 19, 35, 37, 48, 56, 71, 73, 82, 83
United States (USA), 1, 19, 35, 37, 48, 56, 71, 73, 82, 83
USDA, 2, 5, 8, 30, 35, 36, 47, 49, 50, 73, 77, 82, 84

V

valuation, 2, 8, 72, 76
variables, 81
variations, 81, 87
varieties, 32
vegetables, 45
vehicles, 45
venture capital, 42
vertical integration, vii, 1, 17, 18, 29
volatility, 36
vote, 14, 21, 42
voting, 7, 14, 21, 22, 28, 42, 43

W

Washington, 47
water, 34
workers, 20
worldwide, 13, 35

Y

yield, 72